Water and Shore Birds

In the same series

Birds
Wild Flowers
Mushrooms and Fungi
Aquarium Fishes
Minerals and Rocks
Insects
Mammals, Amphibians and Reptiles
Fishes: Freshwater and Marine species
Poisonous Plants and Animals

Chatto Nature Guides

Water and Shore Birds

Illustrated and identified with colour photographs

Dr. Walther Thiede

Translated and edited by
Gwynne Vevers

Chatto & Windus · London

Published by
Chatto & Windus Ltd
40 William IV Street
London WC2N 4DF

*

Clarke, Irwin & Co Ltd
Toronto

British Library Cataloguing in Publication Data
   Thiede, Walther
      Water and shore birds. —
         (Chatto nature guides).
      1. Water-birds — Europe
      I. Title
      598.2'94   QL690.A1

      ISBN 0 7011 2524 1
      ISBN 0 7011 2527 6 Paperback

Title of the original German edition:
Wasservögel · Strandvögel

© BLV Verlagsgesellschaft mbH, München, 1979
English translation © Chatto & Windus Ltd 1980

Printed in Germany

Introduction

This book is intended to interest the reader in water and shore birds, a subject which should appeal to those who live in a country with such an extensive and varied coastline as Britain, and with such a multiplicity of offshore islands, particularly in Scotland. In accordance with their adaptation to an aquatic environment many of the waders, gulls and terns are very similar in coloration and pattern, and this makes severe demands on the observer. He must learn to differentiate, to select the important points at a glance and to be extremely self-critical. With this in mind it is very satisfying to be able, step by step, to learn how to distinguish not only the different species, but also their different plumages, and thus to gain some understanding of their annual cycle. The keen observer will soon learn how to recognize not only the breeding and winter plumages, but also the juvenile plumages at different ages, the differences in some cases between male and female, and even the complication of intermediate or transitional plumages. For it is important to realize that birds do not change their plumage abruptly, but moult step by step in accordance with a hereditary pattern.

The really rare species are not included here and it would not be possible to describe all the different plumages. The aim has, therefore, been to describe the important plumages and other features of the commonest European water and shore birds. Attention has been paid to those plumages which differ only slightly, e.g. the winter plumages of the Common Gull and the larger gulls. The content and selection of the illustrations should primarily be of interest to the informed layman but the professional ornithologist may also find them useful because every extra scrap of information can be of value to identifying water and shore birds. The task of identification in coastal areas is made more difficult by the rapid changes in weather and light, by the distances and by the absence of any fixed point of reference in the vast expanse of a mud-flat or an area of shallow water.

Plenty of time should be allowed for the identification of coastal birds. The observer should lie down in the grass, perhaps in the furrow of a sand dune where he will be well protected from the wind and do nothing but observe. He should turn all his attention to the activities of the birds in the surf and out at sea, not as a duty, nor as an inducement to record his observations, although this too is worth doing, but as a recreation from the trials and tribulations of everyday life.

What are shore and water birds?

These are birds which spend their whole life or a significant part of the year on the water, on the seashore or along the banks of extensive areas of inland water. These include ducks and gulls, terns and herons, cormorants and divers. Like the numerous different waders, which are mostly shore birds, these all show the most diverse adaptations to life in and on the water. They may have webbed feet, the legs may be modified as paddles, the bill may be long and highly sensitive for probing into mud, the legs long and adapted for wading, the wings short and rounded for underwater hunting or long and broad for sailing over the waves for hours or days at a time. Whatever the different adaptations and specializations these birds are all closely associated with water and the shoreline.

Two large birds which feed on fishes have been omitted. These are the Osprey and the Sea Eagle, but they are essentially birds of prey and they are not really adapted for a true aquatic life. In addition certain water birds, such as the Dipper and Grey Wagtail, which live near streams and small rivers have not been included, but they have been described and illustrated in the Chatto Nature Guide entitled *Birds*. Similarly birds such as the Reed Warbler and Marsh Warbler which live close to water were featured in the same volume and are not included here.

In Britain there are many areas of lake and coast where shore and water birds can be observed. On the whole the western coasts have more rocky localities where birds associated with cliffs can be seen.

Bird-watching on the coasts

Watching birds on wide expanses of shore and mud-flat differs in several respects from watching birds in the more restricted confines of a wood or forest.

A pair of human eyes will only see a part of the shore at any one time and the picture seen will differ from that seen by other organisms. When we look straight ahead our vision covers an angle of about 90 degrees, whereas a bird of prey's field of vision is about 180 degrees. The waders have a similar field of vision so that they can see what is behind them without moving their head or eyes. The Jackdaw can also do this. Shore birds can see the bird-watcher much better than he can see them. Birds also differ from humans in colour perception, visual acuity and speed of reaction. The eyes of birds are more efficient and react more quickly than those of humans.

There are also probably differences in the ability to hear. Birds probably do not hear very deep notes while humans cannot pick up very high frequency sounds. Ducks have rather poor hearing.

There are certain other points which affect the success of the bird-watcher in the open expanse of shore and sea.

a) When the tide is coming in the observer should, for instance, be lying well camouflaged in a dry place and allow the birds to come towards him. As the tide rises the area available to them becomes increasingly restricted and they come closer and closer. This is the best time to observe them.

b) Stormy weather is not a disadvantage when watching these birds. At such times they move in closer to the coast, often preferring to move into sheltered bays and inlets. A storm often brings them closer to the land than a rising tide.

c) The sun must not blind you. Birds seen against the sun show no colours, but appear uniformly dark. Always try to watch with the sun behind you.

d) Bird-watchers should walk carefully and without haste. Always walk one behind the other, never in a broad band. Avoid sudden arm movements. On the shore it is better to communicate by calls rather than by using gestures.

e) Keep at a distance from the birds so that they do not fly off prematurely. They can be identified more efficiently when they are standing or walking than when they are in the air. The period during which they are taking off and flying away is nearly always too short for accurate identification.

Identification

The extent and uniformity of the aquatic environment means that the bird-watcher is subject to several sources of error which would not be present if he were observing the birds of a park or wood.

a) There is an increase in contrast when watching birds on a flat expanse of shore. The colours become darker and the birds appeal larger, so that a grey-backed Herring Gull looks like a Great Black-backed Gull.

b) Distances are underestimated. As the distance increases the distinguishing characteristics become less distinct. In such cases the identification should only be of the group, e.g. gulls or diving ducks, without identifying the species.

c) As there are no fixed marks or points of reference the size of the birds is overestimated. They are, in fact, usually smaller than one thinks.

d) It is very common for numbers to be underestimated, for two reasons. First, when the water is choppy the observer at any one time only sees those swimming birds that are on the crest of the waves. In the case of diving ducks a significant number will be underwater. Secondly, when large numbers of birds are densely packed the human eye is incapable of counting them. It is then advisable to count a small section and then to estimate the proportion this bears to the whole group.

e) One's eyes can be blinded, not only by the sun, but also by the water, owing to the reflection of light from its surface. Birds swimming in this reflecting zone cannot be accurately identified. It is also essential to protect one's eyes from the glare and to cease watching and rest for half an hour if there is any trace of headache.

f) The colours of birds alter when the light is too bright or too dim. When the light is no longer sufficient for identification it may still be worthwhile to watch movements and listen to the calls.

Finally, there is no shame in being unable to identify a shore or water bird. This often happens even to the experts.

For those who wish to study the subject in more detail there are several handbooks, and the following may be useful:
A field guide to the birds of Britain and Europe; Peterson, Mountfort and Hollom (Collins)

Birds of Britain and Europe with North Africa and the Middle East; Heinzel, Fitter and Parslow (Collins)
Guide to the identification and ageing of holarctic waders; Prater, Marchant and Vuorinen (British Trust for Ornithology, Beech Grove, Tring)

Field glasses

Swimming birds usually keep at a distance from the land and to view them properly a pair of standard 7×35 binoculars is not usually sufficient. The keen bird-watcher should be equipped with a telescope giving a magnification of 20 to 50 times. This must be mounted on a firm stand to prevent it moving about. With such a telescope it is possible quietly to scan a large area of water and to observe the birds continuously without disturbing them. On scanning an apparently empty lake it is quite astonishing what can be seen with a telescope giving a 20 times magnification. It is then advisable to make a note of what has been seen. If there are two watchers the first can call out what he sees and the second can write it down.

Clothing

In coastal areas it is very easy to underestimate the force of the wind. The result may be severe chilling. Bird-watchers should therefore wear warmer clothing than they normally would. Cotton and wool are the best, with a wind-proof, double-layered cotton anorak worn over several insulating layers. Some sort of head covering is also necessary, even when the sky is overcast and, of course, something to eat and drink.

In a coastal area the bird-watcher should be guided not only by a map but also by his watch. Half tide is the time to start moving back, for distances are deceptive and it is easy to underestimate the dangers of this type of terrain. Storms break out just as suddenly here as they do in the mountains, and it is always advisable to listen carefully to the advice and warnings of the local people.

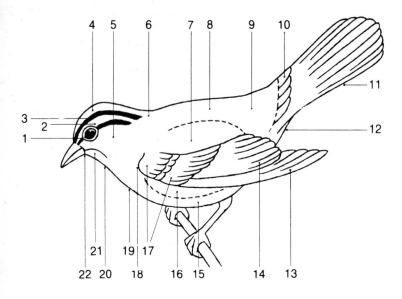

External features of a bird

1	Eye-stripe	12	Under tail-coverts
2	Supercilium	13	Primaries
3	Head-stripe	14	Secondaries
4	Crown-stripe	15	Belly
5	Ear-coverts (cheek)	16	Flanks
6	Nape	17	Wing-coverts
7	Scapulars	18	Bend of wing
8	Back	19	Breast
9	Rump	20	Throat
10	Upper tail-coverts	21	Moustachial stripe
11	Outer tail-feathers	22	Chin

The bird species described and illustrated

The bird species are arranged in the natural or evolutionary order, i.e. with the older families first. In this book the most important groups are the ducks, geese and swans, the waders and the gulls. The following orders are represented:

Gaviiformes	Divers (pp. 14—17) and grebes (pp. 17—21)
Procellariiformes	Petrels (here only the Fulmar, p. 22)
Pelecaniformes	Here only the Gannet and cormorants (pp. 22—25)
Ciconiiformes	Herons (pp. 26—33), storks (pp. 34—35)
Anseriformes	Ducks, geese and swans (pp. 36—69), general remarks (p. 54)
Gruiformes	Rails and cranes (here only the rails, pp. 70—73)
Charadriiformes	
Charadrii	Waders (pp. 74—115), general remarks (p. 74)
Lari	Gulls and terns (pp. 114—137), general remarks (p. 126)
Alcae	Auks (pp. 136—141), general remarks (p. 138)

Water and Shore Birds

Red-throated Diver
Gavia stellata

(fig. above breeding plumage, *below* winter plumage)

The divers of the family Gaviidae breed in the tundra and lakes of the northern hemisphere, and outside the breeding season they live on the sea. The torpedo-shaped body, the paddle-like legs situated far back on the body, the short tail and the dense plumage are all ideally adapted for underwater hunting. On land the divers are very helpless, and move about on their bellies. Four species.—**Characteristics:** the slender, slightly upturned bill is characteristic of this species. During the breeding season the head and sides of the neck are grey, the back of the head has black and white stripes and the throat is red. The back is a dark slate colour. In the winter the plumage is like that of the Black-throated Diver (p. 16) but is paler with fine white speckles. It might be confused with the Black-backed Diver when in breeding plumage as the red throat may appear black in certain lights.—**Distribution**: northern Europe. Breeds in parts of Scotland, and a regular winter visitor elsewhere in Britain.—**Diet**: fishes which are hunted underwater in depths of 2-9 m.—**Breeding:** the nests are at the water's edge, sometimes several together. The clutch of 2 eggs is incubated by both sexes for 26-28 days and the young are fledged in about 6-7 weeks, but the family stay together for longer.

Black-throated Diver

Gavia arctica

Characteristics: the bill is straight and dark. In breeding plumage the sides of the neck have longitudinal black and white stripes, the head is pale grey and the throat black. The upperparts have a checkered pattern of black and white. In the winter plumage the upperparts are a uniform dark grey-brown, the underparts white. This species can possibly be confused with the rarer Great Northern Diver (*Gavia immer*) in which the breeding plumage shows a black head and neck and a striped neck.—**Distribution:** northern Europe. Mainly a winter visitor to Britain, but breeds in small numbers in northern Scotland.—**Diet:** fishes, crustaceans, molluscs, caught at depths of 3.6m.—**Breeding:** the nest is built on islets or on the edges of deep, clean lakes. The clutch of 2 eggs is incubated for 28-30 days, and the chicks are fledged in about 2 months.

Great Crested Grebe

Podiceps cristatus

This is the commonest of the five grebes occurring in Europe. They all live on and in the water. The torpedo-shaped body, the posteriorly positioned legs which act as a rudder, the pointed bill and the very short tail are all adaptations for underwater hunting.—**Characteristics:** in the breeding plumage this species is easily recognized by the red-brown to black frill on the sides of the face. In the greyish-white winter plumage they can be distinguished from the Red-necked Grebe (p. 18) by the pink bill and the white stripe above the eye.-—**Distribution:** Europe, Asia, Africa, in large freshwater lakes. During migration and in the winter it also occurs along coasts and on running waters. In Britain present throughout the year and breeding in many countries.—**Diet:** mainly fishes, water insects, crustaceans, tadpoles.—**Breeding:** pairs remain together for the season. The floating nest is built at the water's edge, preferably in a reed-bed. There are usually 4 eggs which are incubated by both sexes for 25-29 days. The chicks remain with their parents for about 10 weeks. When small the young are carried on the back of one of the parents.

Red-necked Grebe

Podiceps griseigena

Characteristics: in breeding plumage the whitish-grey face contrasts with the deep brownish-red of the neck. The black crown extends down to the red eyes. In the grey and white winter plumage the dark bill has a yellow base and there is no white stripe above the eye. Immature birds have a dark grey-brown face and neck. Not likely to be confused with the other smaller grebes which have a different head shape and a shorter bill.—**Distribution:** shallow inland lakes, moving in winter to the coasts and estuaries. A winter visitor to Britain.—**Diet:** mainly fishes, crustaceans, water insects.—**Breeding:** the nest is built of decaying vegetation. There are usually 4-5 eggs which are incubated by both sexes in turn for 22-23 days. The young remain with the parents for 8-10 weeks.

Slavonian Grebe

Podiceps auritus

Characteristics: a small, more squat grebe with a black head and a tuft of golden feathers extending back from the eyes. The neck is reddish. In winter the upperparts are dark grey, the remainder white. The bill is short and straight. The other grebes are considerably smaller with a different head shape and a shorter bill.—**Distribuion:** shallow lakes in northern Europe. Breeds in northern Scotland, otherwise a regular winter visitor to Britain, especially in coastal areas.—**Diet:** small fishes, insects, crustaceans.—**Breeding:** the nests are built of water weeds in fairly shallow water, sometimes several close to one another. The 3-6 eggs are incubated by both sexes for 20-25 days. The young remain with the parents for 4-6 weeks, and when small they are carried on the back of one of the parents.

Black-necked Grebe

Podiceps nigricollis

Characteristics: a slightly smaller grebe with a black neck when in breeding plumage. The golden feathers radiate from behind the eye. As in the other grebes the winter plumage has greyish-black upperparts and white underparts. The crown is higher and the forehead steeper than in the Slavonian or Red-necked Grebe (p. 18). In winter the dark patch extends below the eye in the present species but not in the Slavonian Grebe. The characteristic upturned bill is not fully developed in juvenile Black-necked Grebes.—**Distribution:** Europe, Asia, North America, in shallow lakes with plenty of vegetation, often moving in winter to coastal areas. Breeds in a few scattered places in Britain, and is common as a winter visitor, particularly in southern England.—**Diet:** mainly insects, also molluscs, crustaceans, fishes.—**Breeding:** nests in shallow water, often in colonies. The 3-4 eggs are laid in a nest of decaying plants and incubated by both parents for 20-21 days. The young dive well at 18 days, but remain for a further period with the parents.

Little Grebe

Podiceps ruficollis

Characteristics: the smallest European grebe, with a squat body, short neck and thick bill. In breeding plumage the upperparts are dark brown, the underparts pale grey-brown; the cheeks, the sides of the neck and the throat are chestnut-brown. The whitish or yellowish-green marking at the base of the bill is characteristic of the species. In winter the plumage is much duller, with the upperparts dark brown, the underparts greyish-white; the cheeks, neck and flanks are yellowish-brown. Might be confused in winter with the Black-necked Grebe, but the Little Grebe is smaller with a stouter and straighter bill.—**Distribution:** Europe, Asia, Africa, on all types of water, provided there are shallow, muddy areas, with plenty of vegetation. A resident breeding bird in most parts of Britain.—**Diet:** insects, crustaceans, tadpoles, in winter small fishes.—**Breeding:** the nest is built of water plants, often sheltered by overhanging branches. The clutch of 4-6 eggs is incubated by both parents for about 20 days. Both sexes tend the young for 6-7 weeks, sometimes carrying them on their backs. Usually two broods a year.

Fulmar

Fulmarus glacialis

This is one of the petrels which are ocean birds that spend their lives out at sea except during the breeding season. They have elongated, tubular nostrils on top of the beak, they breed in colonies and the female lays a single egg.—**Characteristics:** a large, white, gull-like bird with a short neck, long rather narrow wings and a characteristic bill. The wings are a uniform blue-grey and the underparts white. On land Fulmars walk on their heels. They might be confused with the larger gulls but the tubular bill distinguishes them. Also Fulmars do not have black wing tips.—**Distribution:** North Atlantic, North Sea.—**Diet:** crustaceans, squids, fishes and offal taken from the sea surface.—**Breeding:** sexually mature at 6 years old. Nests in colonies on cliffs, occasionally on buildings. The single egg is incubated by both sexes in turn for 49-53 days and the young are fledged at about 50 days.

Gannet

Sula bassana

The gannets and boobies (9 species) are large, powerful marine birds which dive vertically for their prey.—**Characteristics:** large pointed bill, cigar-shaped body, pointed tail. Gannets have five different plumages and are sexually mature at five years. Adults are snow-white with black-tipped wings, a pale yellow head and neck and pale grey eyes. Young in their 1st year have slate-brown, spotted underparts and whitish underparts; in their 2nd year the head, neck and underparts are white, the upperparts blackish-brown; in the 3rd year spotted brown and white above; in the 4th year almost like the adults but with a few dark feathers on the back, wings and tail. —**Distribution:** North Atlantic, North Sea.—**Diet:** fishes.—**Breeding:** nests in large colonies on the ledges of high rocky islands. The nest is a clumsy structure built of seaweed. The single egg is incubated by both sexes in turn for 44 days and the young are fledged at about 3 months.

Cormorant

Phalacrocorax carbo

Cormorants are skilled underwater hunters which swim using the feet only, while steering with the long tail. The plumage is evidently not very well waterproofed and they can be seen drying themselves after each underwater hunt.—**Characteristics:** in breeding plumage (January-May) the body is black with bronze iridescence, a white patch on the thighs and whitish plumes at the back of the head. The throat and cheeks are white. In the winter plumage there are no plumes on the back of the head and no white on the thighs. The juvenile plumage is blackish-brown with white underparts (very variable), but older juveniles have a brown belly. The naked throat is yellowish.—**Distribution:** Europe, Asia, North America, along coasts and on large lakes and rivers. Breeds in Britain.—**Diet:** fishes, some crustaceans.—**Breeding:** sexually mature at 3 years. Breeds in colonies. The nest is built usually of seaweed and the 3-4 eggs are incubated by both sexes for 26-30 days. The young are fledged at about 7 weeks.

Shag

Phalacrocorax aristotelis

Characteristics: a smaller, more slender cormorant with a more slender bill and black plumage showing green iridescence. In the breeding season with a crest but no white. The base of the bill is yellow. The winter plumage is brown with some white on the chin. Juveniles are dark brown with a whitish chin.—**Distribution:** Europe, on rocky coasts. Breeds in many parts of Britain.—**Diet:** marine fishes.—**Breeding:** monogamous, possibly for life, but up to 5% are bigamous when there is a shortage of nesting sites. Sexually mature at 3-4 years. Breeds in colonies on cliffs. The nests are built of seaweed and the 2-5 eggs are incubated by both sexes in turn for 30-31 days. The young are fledged at about 6-7 weeks.

Heron

Ardea cinerea

Herons are long-legged birds with the neck held in an S-shaped curve, loose plumage and a dagger-like bill.—**Characteristics:** pale grey and white with some black markings. In flight the neck is not extended but remains in the S curve. The bill is pale yellow. Juveniles have a blackish bill, a grey instead of a white crown and they lack the two long black crest feathers.—**Distribution:** Europe, Asia, in areas of shallow water along the coasts and on inland waters. In late summer Herons also hunt mice on land.—**Diet:** primarily fishes which are speared at the water's edge, also frogs, mice, insects.—**Breeding:** usually sexually mature in the 3rd year. Breeds in colonies, usually in tall trees. The nest is built of sticks and dead plants, and the 3-5 eggs are incubated by both sexes for 25-28 days. Both parents feed the young which are fledged at 50-55 days.

Purple Heron

Ardea purpurea

Characteristics: a dark grey heron with a rust-brown head and breast and black flanks and belly. The pendant crest feathers are reddish-brown. Juveniles are brownish-red above, pale brown below. At two years the plumage is similar to that of the adult, but with the wing coverts edged with reddish brown and with paler underparts. In flight the neck appears to hang lower than in the Heron and the feet appear larger.—**Distribution:** Europe, Asia, Africa. A vagrant in Britain, usually seen in April-October. More retiring than the Heron.—**Diet:** fishes, frogs, snakes, newts, mice, insects.—**Breeding:** in colonies, the nest being built of reed stems. The 4-5 eggs are incubated by both sexes for about 26 days and the young birds are fully fledged at 6-8 weeks.

Little Egret

Egretta garzetta

Characteristics: a medium-sized snow-white bird with black legs, yellowish feet and a black bill. In breeding plumage the rear of the head and the shoulders carry an elegant pendant crest. Juveniles resemble adults in winter plumage. —**Distribution:** Europe, Asia, Africa, Australia. Appears annually as a vagrant in Britain, mainly in spring and summer. In tropical areas often seen in paddy-fields.—**Diet:** water insects, snails, frogs, lizards, crustaceans, small fishes (up to 15 cm).—**Breeding:** nests in colonies in trees and bushes. The clutch of 3-6 eggs is incubated by both sexes for 21-22 days. The young remain in the nest for 30 days and are independent after a further 2-3 weeks.

Great White Heron

Casmerodius albus

Characteristics: a large, snow-white heron with black feet. The breeding plumage has long plumes on the back, but not on the head. The bill is yellow, but just before the start of the breeding season it is black with a yellow base. Juveniles resemble adults in winter plumage.—**Distribution:** cosmopolitan, in warmer areas. A rare vagrant in Britain.—**Diet:** fishes, insects, small mammals which are taken in shallow waters and paddy-fields.—**Breeding:** nests in colonies, often in reed-beds or areas of similar vegetation. The 3-5 eggs are incubated by both sexes probably for 25-26 days and the young remain in the nest for about 6 weeks.

Squacco Heron

Ardeola ralloides

Characteristics: a squat, crow-sized heron with fawn-coloured plumage and white wings. In the breeding season the bill is characteristically grey with a black tip, in winter it is greenish with a dark tip. In the breeding period the legs are reddish, at other times pale olive-green. The juveniles resemble adults in winter plumage.—**Distribution:** southern Europe, Asia and parts of Africa, in wet areas. Recorded as a rare vagrant in a few places in southern England.—**Diet:** mainly insects, but also frogs, worms, molluscs and small fishes (up to 10 cm).—**Breeding:** often in colonies, sometimes with other herons. The nests are built in trees, bushes or reed-beds. The clutch of 4-6 eggs is incubated by both sexes for 22-24 days and the young remain in the nest for 32-35 days.

Night Heron   (juvenile *on left*, adult *on right*)

Nycticorax nycticorax

Characteristics: a squat, short-legged heron with black upperparts, ash-grey wings and tail, and white underparts. There are three long, white, filament-like feathers on the nape. The powerful bill is blackish-green and the legs are yellow. The juveniles are quite different, being brown with pale markings.—**Distribution:** Europe, Asia, Africa, America, living in wet areas and active at night. A rare vagrant in Britain.—**Diet:** mainly fishes, also newts, frogs, crustaceans and insects.—**Breeding:** usually sexually mature at 2-3 years. A colonial breeder which nests in trees, more rarely among reeds. The 3-5 eggs are incubated by both parents in turn for 21-22 days. The young leave the nest at 3-4 weeks and are independent at 6 weeks.

Little Bittern

(*fig. above left*)

Ixobrychus minutus

Characteristics: a small, pigeon-sized bird which lives a secretive life among reeds, where it becomes active in the evening. The male is pale-buff coloured with a black crown and back, the female yellowish-brown with dark brown streaks on the upperparts. In the male the yellowish-white wing-coverts form a well-defined area which is particularly noticeable when the bird is flying. The bill is yellowish and the legs green. When alarmed the bird extends its neck and assumes a reed-like protective attitude.—**Distribution:** Europe, western Asia, Africa, Australia. Occurs as a vagrant in Britain, mainly in April-June and August-October, but has possibly bred in East Anglia. It is mostly active at dawn and dusk.—**Diet:** fishes, frogs, leeches, insects.—**Breeding:** sexually mature at 2 years. The nest is a slight construction of dead reeds and other vegetation. The 5-6 eggs are incubated by both parents for 17-20 days. The young leave the nest at 17-18 days, perhaps less, and can fly when 30 days old.

Bittern

(*figs above right, and below*)

Botaurus stellaris

Characteristics: a large brown bird with a long neck which is extended vertically upwards when alarmed. The pale yellow plumage with dark brown streaks provides good protective coloration among old reeds. The bill and legs are green. Bitterns fly slowly, rather like an owl, but with the legs trailing behind.—**Distribution:** Europe, Asia, Africa, Australia. In most parts of Britain this is a winter visitor, but it breeds regularly in East Anglia, sometimes also in Kent and Lincolnshire, and possibly elsewhere.—**Diet:** fishes, crustaceans, frogs, newts, insects, small birds and mammals.—**Breeding:** nests in reed-beds, usually in very dense parts. The 4-6 eggs are incubated by the female alone for 25-26 days. The young leave the nest after about 4 weeks, but are not fully fledged until they are about 2 months old.

White Stork
Ciconia ciconia

Storks are long-legged birds with a long, fairly straight neck and a dagger-like bill. They fly rather slowly with the legs and neck extended but drooping slightly.—**Characteristics:** plumage white with black wings, the bill and legs red. Juveniles at first have a black bill, which later becomes brown.—**Distribution:** Europe, Asia, Africa. A rare vagrant in Britain, mainly seen in southern England and East Anglia. Lives in damp meadows and swamp. White Storks from Europe spend the winter in Africa.—**Diet:** frogs, mice, crustaceans, lizards, snakes, fishes, insects, earthworms. —**Breeding:** sexually mature at 3 years. Nests in trees or on buildings. The clutch of 3-5 eggs is incubated by both parents for 33-34 days. The young remain in the nest for 54-63 days.

Spoonbill
Platalea leucorodia

Spoonbills are related to the ibises which have long legs, specialized bills and naked areas on the face or throat.—**Characteristics:** a white bird, almost the size of a White Stork, with a long, black, spoon-like bill which is yellow at the tip in the adult. Breeding birds have long plumes on the nape. The neck and legs are extended in flight. In juveniles the bill is flesh-coloured and the wing tips black.—**Distribution:** Europe, Asia, Africa. In Britain mainly recorded as a vagrant, but it visits East Anglia quite regularly, usually in summer.—**Diet:** small fishes, tadpoles, molluscs, worms, insects. The end of the bill is swept from side to side in the water or mud when searching for food.—**Breeding:** usually nests in colonies, among reeds and other marsh vegetation. The clutch of 2-4 eggs is incubated by both parents for 21-25 days. The young leave the nest after about 4 weeks.

Whooper Swan

Cygnus cygnus

Swans are classified in the same family as ducks and geese (see p. 54).—**Characteristics:** plumage white, bill yellow at the base, black at the tip, without a basal knob. Juveniles are grey with a flesh-coloured bill.—**Distribution:** Europe, Asia, North America. A winter visitor in Britain, particularly to the northern parts. Occasionally breeds in northern and western Scotland.—**Diet:** almost entirely plant matter, but also some worms, water insects and molluscs.—**Breeding:** monogamous. Probably sexually mature at 3-4 years. The nest is built on the ground as a large heap of vegetation. The 5-6 eggs are incubated by the female alone, the male remaining on guard. They hatch in 33-42 days and the cygnets are fledged in about 2 months.

Bewick's Swan

Cygnus columbianus

Characteristics: similar to the Whooper Swan, but more compact and with a shorter, somewhat thicker neck. The front of the yellow patch on the bill is rounded, not pointed as in the Whooper, but this feature varies considerably between individuals.—**Distribution:** Europe, Asia, North America. Recorded as a winter visitor in Britain, mostly from October to April. The subspecies concerned, *Cygnus columbianus bewickii*, breeds in arctic Russia.—**Diet:** water plants and grasses from flooded meadows.—**Breeding:** monogamous. Probably sexually mature at 3-4 years. The 2-5 eggs are laid in a large conical nest on the ground and incubated for 29-30 days. The parents and young remain together during the winter.

Mute Swan

Cygnus olor

Characteristics: a white swan with a long, slender neck which is carried in an elegant S-shaped curve. The base of the bill including the well-defined knob is black, the remainder orange. The knob is larger in the male than in the female. Juveniles have a grey-brown and off-white plumage and a grey bill without a knob. In flight the wing beats can be heard quite clearly.—**Distribution:** Europe, northern Asia. A resident, breeding bird throughout Britain, except the Shetland Islands. —**Diet:** underwater plants plucked from the bottom, also some animal food (small frogs and fishes, worms, molluscs, insects.—**Breeding:** monogamous. Sexually mature at 3 (female) or 4 (male) years. The large nest is built on the ground close to water, and the 5-8 eggs are incubated, mainly by the female, for 35-38 days. The cygnets are fledged in about 4½ months.

Bean Goose

Anser fabalis fabalis

Geese belong to the family Anatidae (see p. 54).—**Characteristics:** a large brownish goose. The bill is black and orange-yellow, the amount of the latter varying considerably. The forehead is relatively flat, and there is sometimes a thin white band at the base of the bill. The legs are orange-yellow. Juveniles have pale legs.—**Distribution:** northern Europe and Asia, wintering further south. In Britain mainly seen as a winter visitor but a few breed in northern Scotland.—**Diet:** grasses, including farm crops, clover, berries.—**Breeding:** monogamous. Probably sexually mature at 3 years. The 4-6 eggs are incubated by the female alone for 27-29 days. The young are tended by both parents for about 40 days. Sometimes known as *Anser arvensis*.

Pink-footed Goose
Anser fabalis brachyrhynchus

Characteristics: a grey goose with a blackish-brown head and neck which contrasts with the paler body and pink legs. The black bill is short and tall, with a pink band. The forehead is relatively high. Juveniles have darker and browner upperparts, and paler, sometimes yellowish, legs.—**Distribution:** breeds in Spitsbergen, and N.E. Greenland. It is a winter visitor to Britain, arriving in September-October and usually leaving in April.—**Diet:** grasses, young wheat, potatoes. —**Breeding:** monogamous. Sexually mature in the 3rd year. Nests on cliffs and in other rocky places. The 3-7 eggs are incubated by the female alone for 26-28 days, the male remaining on guard close by. The young are tended by both parents and are fully fledged at 7-8 weeks. Sometimes regarded as a separate species from the Bean Goose (p. 38) and then known as *Anser brachyrhynchus*

White-fronted Goose
Anser albifrons

Characteristics: a dark-headed grey goose with conspicuous transverse bars on the underparts and a white patch at the base of the bill. The forehead is high and the legs orange. The bill is flesh-coloured but yellow in the Greenland subspecies *flavirostris*. Juveniles do not have the white patch. —**Distribution:** circumpolar. Seen in Britain as a winter visitor from October-January to March-April.—**Diet:** grasses, including farm cereals.—**Breeding:** monogamous. Sexually mature in the 3rd year. The clutch of 5-6 eggs is incubated by the female alone and the young are tended by both parents for about 40 days.

The Lesser White-fronted Goose, *Anser erythropus,* is slightly smaller and the white patch extends on to the forehead.

Grey Lag Goose
Anser anser

Characteristics: a silvery-grey goose. The large bill of the western race is yellow with a pale flesh-coloured tip (known as the nail). The forehead is relatively flat and the legs are flesh-coloured. Juveniles are darker and lack the narrow white ring round the base of the bill.—**Distribution:** Europe, northern Asia. A winter visitor in many parts of Britain. Breeds in small numbers in parts of northern Scotland.—**Diet:** grasses, and other plants.—**Breeding:** monogamous. Sexually mature at the end of the 2nd year. Nests on the ground, the female usually laying 4-6 eggs which she incubates alone for about 28 days. The young are fledged at 10 weeks. Domestic geese are descended from Grey Lags.

Canada Goose
Branta canadensis

Characteristics: a very large grey-brown goose with a black head, neck, bill and legs. The breast and cheeks are white.—**Distribution:** originally Canada and Alaska, but introduced into Sweden, Finland, Norway, Britain and elsewhere. Common in parks.—**Diet:** grasses, clover water plants, in summer also worms, insects and molluscs. —**Breeding:** monogamous. Sexually mature in the 3rd year. Nests on the ground, often on islands, sometimes in colonies. The clutch of 4-7 eggs is incubated by the female alone, but the male remains nearby. The young are fledged at about 2 months.

Brent Goose
Branta bernicla

Characteristics: a small black goose with a white tail and white markings on the sides of the neck. The bill and legs are black. In the dark-breasted subspecies, *B. bernicla bernicla*, the lower breast and flanks are slate-grey, whereas these parts are much paler in the pale-breasted subspecies, *B. bernicla hrota.*—**Distribution:** high arctic, Spitsbergen, East Greenland, arctic islands of Russia. A winter visitor in Britain.—**Diet:** mainly marine vegetation (seaweeds, eelgrass) with some molluscs and crustaceans.—**Breeding:** probably monogamous. Nests on the ground in colonies above the shore zone of lakes and sea coasts. The clutch of 3-5 eggs is incubated by the female for 24-26 days. Fledging time is not known.

Barnacle Goose
Branta leucopsis

Characteristics: a small black and white goose with a short thick head and neck. The crown, neck and breast are black and so is the area just in front of each eye. The remainder of the head is white, the bill and legs black. The upperparts have black and white bars. In juveniles the dark areas of the plumage are grey-brown and the white on the face is not so sharply defined.—**Distribution:** high arctic, Spitsbergen, N.E. Greenland, Novaya Zemlya.—**Diet:** mainly grasses, in winter it feeds particularly on maritime plants.—**Breeding:** monogamous. Probably sexually mature in the 3rd year. Nests in colonies on steep cliffs. The 4-6 eggs are incubated by the female alone and the young are fledged at about 7 weeks.

Shelduck

Tadorna tadorna

Characteristics: a goose-like duck (see p. 54) with blackish-green, white and chestnut areas on the plumage. The legs are flesh-coloured. From July to October the colours are duller and the forehead and cheeks are whitish. The bill is bright red in the breeding birds. The male has a conspicuous knob at the base of the bill, which is absent in the female. Juveniles are dark grey-brown above, with a white face and throat and pale grey legs.—**Distribution:** Europe, Asia, in estuaries and along sandy and muddy coasts, sometimes moving inland to farmland.—**Diet:** mainly insects, molluscs, crustaceans, taken in shallow water, also some vegetable matter.—**Breeding:** sexually mature at 22 months. Nests under rocks or in a burrow, often in a rabbit hole. The 8-12 eggs are incubated by the female alone for 28-30 days. The young are fledged at about 8 weeks.

Wigeon

Anas penelope

Characteristics: legs leaden-grey. From September-October to June the male or drake has a chestnut head with a cream-coloured forehead, a pinkish-brown breast, grey upperparts and flanks and whitish underparts. A broad white area on the wing coverts can be seen when the drake is in flight. In the eclipse plumage (July-October) the drake is similar to the duck but he has white instead of brownish-grey shoulders and a pale grey rather than a slate-grey bill. The female (duck) is reddish-brown with white underparts.—**Distribution:** northern Europe and northern Asia, reaching tropical Africa in winter. Breeds in many parts of Britain, particularly in Scotland.—**Diet:** mainly grasses and eel-grass.—**Breeding:** some breed in their 1st year. Nests on the ground among grasses and heather. The clutch of 7-9 eggs is incubated by the female alone for 22-25 days and the young are fledged at about 6 weeks.

Gadwall

Anas strepera

Characteristics: legs orange-yellow. In breeding plumage (September-June) the drake is pale brown on the head and black on the rear end, with chestnut shoulders and a slate-grey bill. In eclipse plumage (June-September) the drake is similar to the duck but has chestnut shoulders. The female (duck) is brown with a pale brown head and a white belly. Her bill is yellowish at the sides with grey spots. Juveniles are like the female but with brown spots on the underparts. —**Distribution:** Europe, Asia, North America. In Britain mainly recorded as a winter visitor but breeds in East Anglia and in scattered localities in southern England and Scotland. —**Diet:** water plants and some animal food.—**Breeding:** nests on the ground among dense vegetation, close to water. The 7-12 eggs are incubated by the female alone for 24-26 days and the young are fledged at about 7 weeks.

Shoveler

Anas clypeata

Characteristics: the bill is large and spatulate. The legs are orange-red and the shoulders sky-blue. In breeding plumage (October or November to June) the drake has a dark green head, a white breast, chestnut belly and flanks and a dark brown back. In the eclipse plumage (July-September) the drake resembles the duck but has yellow eyes. Females and juveniles have brown eyes.—**Distribution:** Europe, northern Asia, North America. Breeds in many parts of Britain. Birds from Europe spend the winter in Africa.—**Diet:** plant food, including seeds, and crustaceans, molluscs, insects, taken from open water.—**Breeding:** nests on the ground in dry places, but close to water. The clutch of 7-12 eggs is incubated by the duck alone for 22-25 days and the ducklings are fledged at about 6 weeks.

Mallard

Anas platyrhynchos

This is the commonest European duck and is the ancestor of the ordinary farmyard ducks.—**Characteristics:** legs orange-red. In breeding plumage (September-June) the drake has a dark green head separated by a white ring from the dark brown breast. In the eclipse plumage (July-September) the drake is like the female or duck, but is somewhat darker. Juveniles are similar to the females, but have a reddish, horn-coloured bill and yellow to orange-yellow legs.—**Distribution:** Europe, Asia, North America. Breeds very abundantly in Britain.—**Diet:** mostly vegetable matter particularly in winter, but also animal food (insects, molluscs, worms, frogs), mainly in the breeding season.—**Breeding:** nests usually on the ground but sometimes in trees. The 7-11 eggs are incubated by the female alone for about 28 days and the young are fledged in about 7½ weeks.

Pintail

Anas acuta

Characteristics: tail long and pointed. From October to June the drake is grey with the back of the neck and the head brown. The breast is white and there is a characteristic white band extending up the side of the neck. In July-September the drake resembles the female, with pale brown striped plumage but has darker upperparts. In both sexes the bill is leaden-grey and the legs blue-grey.—**Distribution:** northern Europe and Asia, North America, mainly on large lakes, but on migration and in winter is seen along sea coasts and in estuaries.—**Diet:** mostly plant matter, with some insects, worms, molluscs, frogs.—**Breeding:** nests on the ground, often on small islands in lakes or among grasses on sand dunes. The clutch of 7-10 eggs is incubated by the female along for about 23 days. The young, which are tended by the female, are fledged at 6-7 weeks.

Teal

Anas crecca

The smallest European duck.—**Characteristics:** squat body, with dark grey legs and a blackish-grey bill. From October to July the drake has a chestnut head, a green stripe above the eye extending to the nape and a whitish stripe along each flank above the wings. In the eclipse plumage (c. July-October) the drake is mottled brown like the female but with darker upperparts and breast. Juveniles are similar to the female but with spotted brown underparts and a horn-coloured bill.—**Distribution:** northern Europe and Asia, North America, on shallow waters. Breeds in most parts of Britain.—**Diet:** aquatic plants and seeds, worms, insects, molluscs.—**Breeding:** nests on dry ground in sheltered places. The clutch of 8-11 eggs is incubated by the female only for 21-23 days. The female tends the young, assisted by the male. The young are fledged at about 24 days.

Garganey

Anas querquedula

Characteristics: a little larger than the Teal. Bill leaden grey in the male, olive-grey in the female. Legs dark grey. From October to June the drake has a red-brown head with a characteristic broad white stripe over each eye and pale grey flanks; the pale blue fore wings are seen in flight. In eclipse plumage (usually July-September) the drake is mottled brown like the duck, but with blue-grey shoulders. The female has a blackish-brown crown, a broad beige stripe over the eyes and white underparts. Juveniles have brown underparts.—**Distribution:** northern Europe and Asia. European birds winter in Africa. In Britain seen mainly as a summer visitor, but it breeds in some southern counties of England.—**Diet:** water plants, worms, molluscs, young fishes, frogs.—**Breeding:** nests among grasses close to the water. The clutch of 8-11 eggs is incubated by the female only, who also tends the young.

Red-crested Pochard

Netta rufina   (*fig. above* female, *below* male)

Characteristics: a small, compact diving duck with a relatively large head. The drake has a bright red bill, a chestnut head, pale red eyes, and glossy black neck, breast and underparts. The duck is dark brown above, pale brown below, with whitish-grey cheeks. Males in eclipse plumage (July-September) and juveniles are similar to females. —**Distribution:** Europe and central Asia. In Britain recorded as an autumn visitor, mainly in eastern England.—**Diet:** mostly water plants, with a small amount of animal food.—**Breeding:** nests on the ground among dense vegetation, close to water. The clutch of 8-10 eggs is incubated by the female alone for 26-28 days and the young are fledged at about 45 days.

Ducks, geese and swans   (pp. 36-69)

These birds all belong to the Anatidae, a family in the order Anseriformes. They are waterfowl with a specialized bill, webbed feet and dense down. The males have a penis. The bill is furnished with lamellae and in most species is covered by soft skin. When feeding water is sucked in, with the bill slightly open. The bill is then closed and the tongue presses against the lamellae, pushing the water out and leaving the food inside the bill. Mergansers have sharply serrated bills with which they can seize and hold fishes. During the full moult towards the end of the breeding season all species are unable to fly for 3-7 weeks as they moult all their flight feathers at the same time. During this period many species congregate in a protected moulting site where there is plenty of food.

Pochard

Aythya ferina

The diving ducks of the genus *Aythya* have a short compact body with the legs positioned rather far back, so that when standing they assume an oblique position. Unlike the ducks of the genus *Anas* they cannot take off directly from the water, but have to skim along the surface before doing so.—**Characteristics:** from October to July the drake Pochard has a red-brown head and neck, a black breast and tail region and grey back and flanks. The eyes are orange to red, the bill black with a pale grey middle part. In the eclipse plumage (June-October) the black areas become brown and the red-brown is duller. The duck is dark brown with grey back and sides, with fine dark wavy markings. The eyes are brown. Juveniles are a uniform dark brown with a wavy pattern on the sides and wings. It is characteristic of both sexes that there is no white on the wings.—**Distribution:** Europe, northern Asia. In Britain, Pochards can be seen at all times of the year. They breed regularly in many parts of eastern Britain (Orkneys to Kent) and also less frequently in other parts.—**Diet:** mainly water plants, with some animal food (worms, insects, molluscs, frogs).—**Breeding:** nests in or near to the water. The clutch of 6-11 eggs is incubated by the female alone for 24-28 days and the young are fledged at about 50 days.

Ferruginous Duck

Aythya nyroca

Characteristics: a dark mahogany-coloured duck with a conspicuous white area under the tail (in the adults of both sexes). The bill and legs are grey. The eyes are white in the male, dark in the female. Juveniles are like the females, but with brown under the tail.—**Distribution:** Europe, central Asia, northern Africa in quiet lakes and marshes with good cover, but in some places it occurs on the sea during winter. A fairly regular winter visitor to Britain.—**Diet:** water plants, particularly duckweed, and some aquatic insects and worms.—**Breeding:** nests on the ground among shore vegetation, sometimes several together. The clutch of 7-11 eggs is incubated by the duck alone for 25-27 days; the drake is usually in attendance. The young are tended by the duck and are fledged at about 50 days.

Scaup
(fig. left female; right male)

Aythya marila

Characteristics: both sexes have yellow eyes. From October to July the drake has a black head, breast and shoulders, a pale grey back with a vermiculated pattern and white flanks. In the eclipse plumage (July to end of October) the drake is dark brown with the flanks and upperparts pale grey with brown markings. The duck is dark brown with a broad white marking round the base of the greyish bill. Juveniles are similar to the female, but with round, pale beige spots at the base of the bill.—**Distribution:** Europe, northern Asia, North America. In Britain recorded mainly as a winter visitor, but it has bred in northern Scotland and the Outer Hebrides.—**Diet:** mainly animal food (molluscs, crustaceans, small insects) and some water plants.—**Breeding:** probably sexually mature in the 2nd year. The 7-12 eggs are incubated by the duck for 26-28 days. The young are fledged at about 45 days.

Tufted Duck

Aythya fuligula

Characteristics: both sexes have yellow eyes. In winter and summer the male is black with white flanks. The head is iridescent purplish-violet with a tufted crest at the back. The female and juveniles are blackish-brown. In the eclipse plumage the male is brown.—**Distribution:** Europe, northern Asia. In Britain this species breeds in most counties and has fairly recently extended its range to the Shetland Islands. —**Diet:** mainly animal food (molluscs, insects, frogs and spawn, small fishes) with some water plants.—**Breeding:** nests in sheltered places among vegetation close to the water, sometimes several together. The clutch of 6-14 eggs is incubated by the female only for 23-25 days. The young, tended by the female, are fledged at about 6 weeks.

Eider

Somateria mollissima

Characteristics: a large duck with a characteristic broad base to the bill. The male has a striking plumage with a black crown, belly, sides and tail region, the rest of the body being mainly white. The female is brown with blackish streaks and mottling. In eclipse, the male is blackish but with white on the wing.—**Distribution:** Europe, northern Asia, North America, Spitsbergen, Greenland. In Britain, Eiders breed in fairly large numbers in the Hebrides, Orkneys, Shetlands, also along the west coast of Scotland and parts of the east coasts southwards to the Farne Islands. This is essentially a marine duck.—**Diet:** chiefly animal food (insects, molluscs, small fishes) with some plant matter.—**Breeding:** sexually mature by the 3rd year at the latest. Nests on the ground in colonies. The clutch of 4-6 eggs is incubated by the female alone for 26-28 days. The young are fledged at about 9-11 weeks.

Common Scoter
Melanitta nigra

Characteristics: the drake is glossy black, its bill is also black but with a conspicuous patch of yellow. The duck and the juveniles are blackish-brown with pale cheeks and chin, and a grey bill. In eclipse, the male is duller, but retains the yellow on the bill.—**Distribution:** Europe, northern Asia, North America. In Britain seen mainly as a winter visitor to the south and east, but a few breed in northern Scotland and the Inner Hebrides.—**Diet:** mainly molluscs, but also some crustaceans, worms, insects and water plants.—**Breeding:** sexually mature at 2-3 years. Nests on the ground, usually quite close to the water. The clutch of 5-7 eggs is incubated by the female alone for 29-31 days. The young birds are tended by the female and are fledged at about 6-7 weeks.

Velvet Scoter
Melanitta fusca

Characteristics: this is a larger bird than the Common Scoter and it has a distinctive white patch on the wings and a small white area in the region of the eye. Males are black with a yellow edging to the bill and red legs. Females and juveniles have two pale grey patches on the face, one in front of, the other behind the eye, and they have dull reddish legs and a grey bill.—**Distribution:** Europe, northern Asia, North America. In Britain mainly seen as a winter visitor between September and April-May. There is a possibility that it has occasionally bred in Scotland.—**Diet:** mainly molluscs, also crustaceans, worms, insects, and during the breeding season a certain amount of plant food.—**Breeding:** sexually mature in the 2nd year. Nests on the ground among grasses and small bushes. The clutch of 7-10 eggs is incubated by the female alone for 26-29 days. The young are tended by the female for 4-5 weeks and are fledged at about 2 months.

Long-tailed Duck (*fig.* male in breeding plumage)
Clangula hyemalis

Characteristics: a marine duck with a confusing array of plumages in both sexes. The large eyes are pale brown to orange in the male, brown in the female. In winter both sexes have a mainly white head, white belly and flanks, and the male has a distinctive pattern of dark brown and white and a characteristic long tail. Between February and June the male has the head, neck and breast dark brown, the sides of the face white and the back brown. The eclipse plumage in July-August is rather duller and the long tail feathers are moulted. In winter the female is brown above, with white on the sides of the neck.—**Distribution:** Europe, northern Asia, North America. In Britain, mainly a winter visitor, but it has bred in the Orkneys and Shetlands.—**Diet:** molluscs, crustaceans, small fishes, insects.—**Breeding:** sexually mature in the 2nd year. Nests on the ground close to water. The 5-9 eggs are incubated by the female only for 24-26 days. The young are fledged at about 5 weeks.

Goldeneye
Bucephala clangula

Characteristics: a compact diving duck. From October to July the male is black above with an iridescent blackish-green head, and white below. There is a characteristic white patch between the golden-yellow eye and the bill. The legs are orange-yellow and the bill dark grey. The females have a mottled grey plumage with a brownish head, a white collar, a dark grey bill and pale yellow to white eyes. Juveniles are browner, without the collar, and with yellow-brown legs. —**Distribution:** Europe, northern Asia, North America. In Britain seen as a winter visitor and during migration.—**Diet:** molluscs, insects, crustaceans, with some water plants in autumn.—**Breeding:** sexually mature in the 2nd year. Nests in holes, sometimes rabbit burrows. The 8-11 eggs are incubated by the duck alone for 29-30 days and the young are fledged at 8-9 weeks.

Smew

Mergus albellus

This and the following two species are known as sawbilled ducks.—**Characteristics:** bill and legs leaden-grey. From November to June the male is pure white with black on the face and nape, and also on the back. In July-October the male and female are grey with a red-brown crown, white cheeks and throat. The juveniles are brown rather than grey.—**Distribution:** Europe, northern Asia, in standing and slow-flowing waters. In Britain known only as a winter visitor.—**Diet:** mainly animal food, particularly fishes, some water insects.—**Breeding:** probably sexually mature in the 2nd year. Nests in holes in trees. The clutch of 7-9 eggs is incubated by the female only for about 4 weeks.

Red-breasted Merganser

Mergus serrator

Characteristics: a large, slender, sawbilled duck with a thin, pointed bill. The legs and bill are red in the male, yellowish-red in the female and juveniles. From October to July the male has a blackish-green head and crest, red eyes, a white collar, a brown breast and a white belly. Females and juveniles are brown with a white belly.—**Distribution:** Europe, northern Asia, North America. It breeds in many parts of Scotland and in a few places in England.—**Diet:** fishes, with some insects, worms and crustaceans. —**Breeding:** probably sexually mature in the second year. Nests in holes under trees, in shallow burrows, or hidden among vegetation close to the water. The clutch of 6-12 eggs is incubated by the female only for about 30 days. The young from several pairs often go about with a single female.

Goosander

Mergus merganser

(*fig. above* male;
below female with young)

The large sawbilled ducks swim with the body deep in the water and thus have a flattish, streamlined appearance. Their narrow thin bills have sharply serrated edges and a hook-like tip (the nail), both excellent adaptations for catching fishes.

Characteristics: the Goosander is larger than a Mallard and has a blood-red bill. From September to June the male has a white breast and belly with a salmon-pink tinge, a black back, an iridescent greenish-black head and neck, blackish-brown eyes and red legs. From about July to September the male is like the female but with white in the front of the head and neck. The female is pale grey with a rich chestnut-brown head and upper throat, a white breast and underparts. The legs are orange. Juveniles are similar to the female but have shorter crests and browner upperparts; they have a yellow-brown bill and yellow eyes.—**Distribution:** Europe, northern Asia, North America. Breeds in several counties in Scotland, particularly in the north, and in a few places in northern England. Seen elsewhere as a winter visitor.—**Diet:** mainly fishes, with some water insects, crustaceans, frogs. —**Breeding:** sexually mature in the 2nd year. Nests close to the water in hollow trees, in holes in the banks or among boulders. Sometimes several pairs nest close together. The clutch of 7-13 eggs is incubated by the female alone for 30-35 days. The young leave the nest after 2-3 weeks and are tended by the female, who sometimes carries them on her back, until they are fledged at 60-70 days.

Water Rail

Rallus aquaticus

Rails are diurnal birds adapted for living in wet habitats with dense vegetation. The body is much compressed laterally, the backbone is very mobile and the front toes very long. The tail is short and the wings rounded.—**Characteristics:** about the size of a Blackbird, with a long red bill. The face and underparts are slate-grey, the flanks barred black and white, and the upperparts olive-brown with black streaks. Juveniles have brownish underparts. The bird takes to the wing unwillingly.—**Distribution:** Europe, northern Asia. In Britain, breeds in several marshy areas, particularly in Norfolk, but less abundantly in Scotland.—**Diet:** crustaceans, molluscs, water insects, small fishes.—**Breeding:** sexually mature in the 2nd year. Nests on the ground in a sheltered place, usually close to the water or among reeds. The 6-11 eggs are incubated by both sexes for 19-21 days and the young are fledged at 7-8 weeks. There are 2 broods a year.

Spotted Crake

Porzana porzana

Characteristics: a rail about the size of a Blackbird which, in contrast to the Water Rail, has a short, yellowish or olive-green bill and green legs. The base of the bill is red in the male, orange-yellow in the female. The breast is olive-grey with white dots, the belly grey-brown and the flanks pale olive-brown with cross-banding.—**Distribution:** temperate Europe and Asia, wintering further south. An extremely secretive bird living in swamps, marshes and ponds.—**Diet:** molluscs, water insects and their larvae, soft plants and seeds.—**Breeding:** nests on the ground among sedges and grasses. The clutch of 8-12 eggs is incubated by both sexes for 18-20 days. Both sexes tend the young which are fledged at 35-42 days.

Moorhen

Gallinula chloropus

A rail which in many areas is a common bird in city parks.—**Characteristics:** bill and frontal shield red. The plumage is greyish-black with thin, oblique flank stripes. The under surface of the tail is white and the legs green. Juveniles are grey-brown with a beige throat.—**Distribution:** almost cosmopolitan, in marshy areas and on small inland lakes. A common resident bird, breeding throughout Britain.—**Diet:** chiefly plant food, such as leaves, grasses and seeds, but also worms, water insects, snails, tadpoles and fishes.—**Breeding:** nests on the ground usually among water plants or bushes, more rarely in trees. The clutch of 5-11 eggs is incubated by both sexes for 17-24 days and the young are fledged at 6-7 weeks. Usually 2 broods a year, sometimes 3.

Coot

Fulica atra

A rail adapted for diving and often seen on lakes in cities.—**Characteristics:** a black bird with a white bill and frontal shield. The lobed feet vary in colour from leaden-grey to yellow. The eyes are red in the adults, brown in the juveniles, which are similar to the juvenile Moorhens, but without the white area under the tail.—**Distribution:** Europe, Asia, Australia. A common resident breeding bird throughout Britain.—**Diet:** mainly plant food, particularly the soft shoots of water plants, together with molluscs, insects, worms, tadpoles and fishes.—**Breeding:** nests among dense vegetation close to the water's edge. The clutch of 6-9 eggs is incubated by both sexes for 22-24 days and the young are fledged at about 8 weeks. There may be 2 or even 3 broods a year.

Waders or Charadrii (pp. 74-115)

This sub-order constitutes the largest single group (35 species) among the 144 species described in this book. The order Charadriiformes, which also includes the gulls, terns and auks, here comprises 58 species.

The waders are long-legged ground birds, mostly with brownish plumage. The bill is usually long and well adapted for probing mud. They fly rapidly and vary in size from a sparrow to a domestic fowl. The eggs are laid on the ground. They are all nidifugous, i.e. the young leave the nest almost immediately after hatching.

The five families dealt with are the oystercatchers (Haematopodidae), the plovers (Charadriidae), the snipe (Scolopacidae), the avocet and stilts (Recurvirostridae) and stone curlews (Burhinidae).

Lapwing
Vanellus vanellus

Among the numerous waders the Lapwing is one of the most abundant inland breeders and the one most likely to be seen by the public.

Characteristics: a black and white bird, about the size of a pigeon, with broad, rounded wings, which is immediately recognizable by the characteristic display flight and the 'pee-weet' call (hence the Scottish name Peewit for this bird). The dark plumage shows purple and green iridescence. —**Distribution:** Europe, northern Asia, mostly in flat country with low vegetation. A common resident breeding bird in all parts of Britain.—**Diet:** chiefly insects, but also worms, molluscs, crustaceans, and some vegetable matter.—**Breeding:** nests in a shallow muddy hollow lined with grasses. The 4 eggs are incubated by both sexes for 24-29 days and the young are fledged at 30-42 days.

Ringed Plover
Charadrius hiaticula

Characteristics: distinguished from the very similar Little Ringed Plover (see below) by the slightly larger size and the conspicuous white wing bar seen when in flight. In breeding plumage the bill is orange-yellow with a black tip. In winter plumage the black face markings and breast band are browner and the bill is black. The legs are orange-red (in breeding plumage), or brownish-red (in winter). Juveniles lack the black face markings and have a pale to dark brown breast band.—**Distribution:** Europe, northern Asia, North America. Breeds in many parts of Britain.—**Diet:** worms, molluscs, crustaceans, insects and some plant food.—**Breeding:** nests on sand or gravel, sometimes on turf. The clutch of 4 eggs is incubated by both sexes for about 25 days and the young are fledged at about 25 days. There are 2, sometimes 3, broods a year.

Little Ringed Plover
Charadrius dubius

Characteristics: upperparts sandy brown, underparts white. Somewhat smaller than the Ringed Plover from which it differs in lacking the white wing bar and in having a black bill and a yellow ring round each eye. The legs are pale yellowish. Juveniles have brownish-yellow legs, a yellowish-grey ring round the eye, and reddish-yellow edges to the feathers on the upperparts.—**Distribution:** Europe, Asia. In Britain now a regular summer visitor, mainly in the south-east, where small numbers are breeding. Up to about 40 years ago this bird was only recorded as a rare vagrant in Britain.—**Diet:** mainly insects, but also spiders, molluscs and worms.—**Breeding:** nests on sand, shingle or among grass. The clutch of 4 eggs is incubated by both sexes for 22-28 days and the young fly at 24-30 days.

Kentish Plover
Charadrius alexandrinus

Characteristics: upperparts sandy-coloured, underparts white. A small plover with a black bill, black legs, a white wing bar and a less extensive black area on each side of the upper breast. The male has a black mark above the white forehead but this is lacking in the female. Juveniles are like the female but have grey legs and dark brown or reddish-yellow edges to the feathers on the upperparts.—**Distribution:** Europe, Asia, North America. In Britain, mainly seen when on migration in spring and autumn. It has bred in Kent, Sussex and East Anglia but no longer does so regularly.—**Diet:** chiefly insects, worms, molluscs and crustaceans.—**Breeding:** nests on the ground on shingle with scattered grasses. There are usually 3 eggs which are incubated by both sexes for 23-28 days. The young are fledged at about 6 weeks.

Dotterel
Eudromias morinellus

Characteristics: about the size of a thrush with conspicuous broad white stripes above the eyes and a narrow white band separating the grey-brown upper breast from the chestnut lower breast. The crown and upperparts are dark brown, the cheeks whitish and the belly black. The legs are yellowish and the bill black. Females are usually more brightly coloured than males. Juveniles are dark brown above with pale yellow edges to the feathers, and the characteristics broad eye stripes are beige.—**Distribution:** Europe, northern Asia, on mountains during the breeding season. A summer visitor in Britain where a few pairs breed in the Grampians, Cairngorms and eastern Ross.—**Diet:** chiefly insects and spiders, but on migration some molluscs and plant food.—**Breeding:** nests on the ground. There are usually 3 eggs which are incubated almost or entirely by the male only for 24-28 days. The young are tended mainly by the male and are fledged in about a month.

Grey Plover

Pluvialis squatarola

(fig. winter plumage)

Characteristics: in winter the adults have brownish-grey upperparts, which are obscurely mottled. In summer the upperparts have silver-grey spangling. Juveniles have the upperparts more yellowish and are then very similar to juvenile Golden Plovers. At all ages, however, the axillaries (feathers close to the body under the wings) are black. The bill is black and the legs grey.—**Distribution:** Europe, Asia, America, Australia. In Britain, mainly seen as a winter visitor and when migrating in autumn and spring.—**Diet:** insects, crustaceans, molluscs, worms and some berries.—**Breeding:** sexually mature at 2 years. Nests on the ground. The clutch of 4 eggs is incubated by both sexes for about 26 days, but the time of fledging is apparently now known.

Golden Plover

Pluvialis apricaria

(fig. left winter plumage,
right summer plumage

Characteristics: similar to the Grey Plover but with white axillaries (*see above*) and at all times the upperparts are spangled with black and gold. In summer the face and underparts are black and there is a broad white area running down from the forehead, over and behind the eye and down the sides of the neck and breast. The legs are greenish-grey.—**Distribution:** Europe, northern Asia. In Britain breeds throughout Scotland and in northern England and Wales.—**Diet:** insects, worms, spiders, crustaceans, snails with some berries and grasses.—**Breeding:** sexually mature in the 2nd year. Nests on the ground among short vegetation. The clutch, usually of 4 eggs, is incubated by both sexes for about 28 days and the young are fledged at about 32 days.

Oystercatcher

Haematopus ostralegus

Characteristics: roughly the same shape as a Lapwing, but with black and white plumage, a large orange-red bill and pink legs. In winter the throat has a white half collar. Juveniles have mottled brown upperparts and a dark brown bill.—**Distribution:** Europe, Asia, America, Australasia. Breeds abundantly in Scotland on the coast and also inland, and less commonly in northern England and Wales.—**Diet:** mainly cockles and mussels, but also snails, crustaceans, worms and insects.—**Breeding:** sexually mature in the 3rd-5th year. Nests in a shallow depression in the ground lined with shells and pebbles. There are normally 3 eggs which are incubated by both parents for 26-28 days. The young are fledged at 34-36 days.

Turnstone

Arenaria interpres

Characteristics: in the breeding season the head and underside are white, the facial mask and breast black. The wings and back have a coarse pattern of black and pale red-brown, and the legs are orange-red. In the winter the head, upperparts and sides of the head are coffee-brown with broad white edges to the feathers, the legs are pale orange-red, the underparts white and the breast blackish-brown. Juveniles have brownish-yellow to pale orange legs.—**Distribution:** Europe, northern Asia, North America. Recorded in Britain as a winter visitor and also when on migration in autumn and spring.—**Diet:** molluscs, insects, crustaceans.—**Breeding:** sexually mature in the 2nd year. Nests on the ground. The clutch of 4 eggs is incubated by both sexes for 21-23 days and the young are fledged at 24-26 days.

Snipe
Gallinago gallinago

Characteristics: a long-billed brown bird about the size of a thrush which only takes to the wing at the last minute, flying off in a zigzag course. The wings are a uniform dark brown and the belly white without a pattern. The pale stripes on the crown run longitudinally, not transversely as in the Woodcock *(Scolopax rusticola)*. The drumming or bleating flight occurs particularly during the courtship period, but also at other times; the sound produced is due to the vibration of the outer tail feathers.—**Distribution:** Europe, Asia, North America, in marshland, ponds and wet meadows. Breeds in many parts of Britain.—**Diet:** chiefly worms, taken by probing in mud with the sensitive bill, also molluscs, insects, and some seeds and grass.—**Breeding:** nests in a well-concealed place among grasses or rushes. There are usually 4 eggs which are incubated by the female only for 18-20 days. Both parents tend the young for about 4 weeks.

Jack Snipe
Lymnocryptes minimus

Characteristics: a compact bird, rather smaller than a Snipe with a relatively shorter bill. Flies silently and somewhat like a bat, but does not zigzag. The blackish-brown crown lacks a pale stripe. The almost black back has 2 broad, yellow longitudinal stripes.—**Distribution:** Europe, northern Asia, on moorland in marshy country and wet meadows. A winter visitor to Britain.—**Diet:** mainly worms, insects and molluscs, but seeds are also taken.—**Breeding:** nests among sedges and grasses, sometimes in low birch scrub. The clutch of 4 or occasionally only 3 eggs is evidently incubated by the female alone for about 24 days, but the time of fledging is not known.

Curlew

Numenius arquata

Characteristics: a large brown bird with long greenish-grey legs and a long bill which is curved downwards. The plumage is striped yellow-brown and dark brown, with paler un-derparts. The Curlew differs from the very similar Whimbrel (see below) in its greater size, in lacking the conspicuous stripes on the crown and in having a different call. Also the bill is longer than in the Whimbrel. There is not much dif-ference between the summer and winter plumages, except that the latter is usually paler. Juveniles are similar to the adults in summer, but with a shorter bill and warmer brown un-derparts.—**Distribution:** Europe, northern Asia, in moorland and heathland, at certain times on coasts and in estuaries. A resident bird in Britain, breeding in nearly all parts of Scotland, in all English counties north of Derbyshire and also in Wales, Devon and Cornwall.—**Diet:** chiefly insects, molluscs, worms and frogs when living inland, also berries and seeds. On the coasts it eats crustaceans, marine worms and small fishes.—**Breeding:** nests on the ground in a shallow depression lined with grasses. There are normally 4 eggs in a clutch, but sometimes up to 6. These are incubated by both sexes for about 30 days. The young are tended by both parents and are fledged at 5-6 weeks. In some parts of their range Curlews are threatened by the drainage of the type of country which forms their natural habitat.

Whimbrel

Numenius phaeopus

Characteristics: a smaller bird than the Curlew with a relatively shorter bill and a different call. The two broad dark brown stripes on the head, separated by a pale stripe, also distinguish this species from the Curlew.—**Distribution:** Europe, northern Asia, North America, on heaths and grasslands, but not in really wet terrain. In Britain seen mainly when on migration in April-June and July-October. It does, however, breed regularly in the Shetland Islands and has also bred in the Orkneys and parts of northern Scotland.—**Diet:** insects, worms, molluscs, crustaceans, and also some berries.—**Breeding:** nests on the ground in a shallow depression lined with a little dry grass and moss. The clutch of usually 4 eggs is incubated by both parents for 27-28 days. The young are fledged at 5-6 weeks.

Black-tailed Godwit

Limosa limosa

Characteristics: a long-legged and long-billed wader, somewhat larger than a Lapwing. Other distinguishing features include the white tail and the broad white wing bars which are seen when the bird is in flight. In the breeding plumage (February-August) the neck and breast are rust-brown, the flanks banded and the belly whitish. The female often shows less rust-brown and fewer bands on the flanks. In the winter plumage the head, neck and breast are pale grey, without any rust-brown, the back is dark grey and the flanks white.—**Distribution:** Europe, northern Asia. In Britain mostly seen as a regular visitor, particularly in July-September and March-May. In recent years it has, however, bred in northern Scotland, Norfolk and Lincolnshire.—**Diet:** insects, crustaceans, molluscs, worms.—**Breeding:** nests on the ground in a shallow depression lined with dead grasses. The 4 eggs are incubated by both parents for 22-24 days. The young are fledged at about 35 days.

Bar-tailed Godwit

Limosa lapponica

(*fig. above* winter plumage, *below* breeding plumage)

Characteristics: slightly smaller than the Black-tailed Godwit (p. 88) but otherwise very similar. The most important distinguishing features are the white tail with a broad black bar, the slightly upturned bill and the absence of white wing bars. Also when in flight the legs do not project so far beyond the tail as they do in the Black-tailed Godwit. In the breeding plumage (April to August or September) the male has rust-brown underparts while the female is a paler brown, with mainly white underparts. In winter plumage (August or September to April) the adults are grey-brown above with pale rust-coloured feather edges, and white below. Juveniles are similar to the adults in winter plumage but more yellow-grey.—**Distribution:** Europe, northern Asia, Alaska, on wet lands, sewage farms, the edges of lakes and in winter in estuaries and on sandy and muddy shores. In Britain seen mainly as a winter visitor and when on migration, but it does not breed.—**Diet:** on its breeding sites mainly beetles, mosquito larvae and other insects, but on the coasts it feeds largely on marine worms, molluscs and crustaceans. —**Breeding:** probably sexually mature in the 2nd or 3rd year. Nests on the ground in a hollow lined with dry birch leaves and lichens. The clutch, which usually has 4 eggs, sometimes only 2 or 3, is incubated by both sexes probably for about 3 weeks.

Spotted Redshank

Tringa erythropus

(*fig. above* winter plumage, *below* breeding plumage)

Characteristics: a large slender wader which is mainly black in the breeding plumage, pale grey above and whitish below in winter. The long legs are dark red in the breeding season, red in the winter. The long, thin bill is black, with the base of the lower mandible red. The winter plumage is not unlike that of the Redshank (p. 94), but this is a slightly larger, greyer bird with barred and spotted wing coverts. Juveniles are darker and browner in the winter and their feathers have pale edges.—**Distribution:** northern Europe and northern Asia, on marshes, lakes and reservoirs, moving in winter to coastal areas, including saltmarshes. In Britain it is seen mostly in winter, but also at other times of the year and particularly during the periods of migration. It is scarcely ever seen in northern Scotland.—**Diet:** chiefly insects, but also worms, crustaceans, molluscs taken in shallow water. The bird sometimes catches frogs and small fishes.—**Breeding:** the age of sexual maturity is evidently not known. Nests on the ground, in a slight depression with a small amount of grass and dead leaves as a lining. The clutch of 4 eggs is incubated for an unknown period, probably mainly if not entirely by the male. After hatching the young are at first tended by both parents, but the female soon moves away. There is a single brood in the year.

Redshank

Tringa totanus

Characteristics: an active brown wader with orange-red legs and an orange bill with a dark tip. In flight the broad white area at the rear of each wing and the white rump are quite characteristic. In summer the basic colour is a warm brown, the back, head and neck having black streaks. In winter, the head, neck and breast are greyish with dark streaks and the back is ash-brown. Juveniles are similar to this but the bill is pale yellowish-red with a black tip and the legs are pale red.—**Distribution:** Europe, northern and central Asia, in open country with water, and in wetlands. In Britain the Redshank is seen at all times of the year and it breeds in all parts of England (except the extreme south-west) and also throughout Scotland. The resident population is augmented by migrants from parts of continental Europe and from Iceland.—**Diet:** insects, crustaceans, worms, molluscs and some small fishes, together with a certain amount of leaves, buds, seeds and berries.—**Breeding:** nests on the ground in a well-hidden place, often among tall grasses, so that the eggs are not visible from above. The clutch, which usually has 4 eggs, is incubated by both sexes for 22-25 days, and the young leave the nest very soon after hatching. They are tended at first by both parents but the female leaves before the young are fledged at 27-35 days.

Greenshank

Tringa nebularia

Characteristics: an active, slender, long-legged wader about the size of a Spotted Redshank (p. 92). The long, greyish bill is slightly upturned and the legs are olive-green to yellow-green. The wings are a uniform dark grey-brown without wing bands. The rear part of the back and the rump are white. In the summer plumage the upperparts are brownish-grey to blackish-brown, the head, throat and breast have dark brown spots or streaks and the underparts are white. In the winter the upperparts are grey, the underparts white with fine streaks on the breast.—**Distribution:** Europe and northern Asia. In Britain mainly a summer and winter visitor, but it breeds in parts of northern Scotland, including some of the islands in the west.—**Diet:** insects, worms, molluscs, crustaceans, small fishes.—**Breeding:** nests on the ground. The clutch of 4 eggs is incubated by both parents for 23-25 days and the young are fledged at about a month.

Common Sandpiper

Actitis hypoleucos

Characteristics: a small, short-legged wader, about the size of a lark, often seen inland on lakes and streams. The upperparts are dark brown (the feathers having blackish-brown edges), the throat and breast grey-brown. The rear part of the body is jerked up and down constantly and this displays the white sides of the tail. The bird flies off low over the water when disturbed, giving a shrill call. The bill is dark horn-coloured and the legs grey-green to yellow-green.—**Distribution:** Europe (except Iceland) and northern Asia, close to streams and lakes during the breeding season. Breeds in Scotland and northern England as far south as Herefordshire, and more locally in Devon and Somerset.—**Diet:** insects, water spiders, worms, crustaceans and some vegetable matter.—**Breeding:** nests on the ground close to water but usually in a place sheltered by grasses. The clutch of 4 eggs is incubated by both sexes for 21-22 days and the young are fledged at about 4 weeks.

Green Sandpiper

Tringa ochropus

(fig. above)

Wood Sandpiper

Tringa glareola

(fig. below)

These two species are among the most difficult birds to distinguish.—**Characteristics:** two waders about the size of a Starling, with white rump, white underparts, dark brown upperparts and a blackish bill with an olive-coloured base. The Green Sandpiper, which is mainly blackish-brown and brilliant white, has leaden-grey legs with greenish joints, only 3-4 blackish-brown cross bars on the tail and blackish undersides to the wings. In the breeding plumage the rows of fine pale dots are most distinct on the upperparts and there is a black streak in front of each eye. The legs are relatively shorter, the bill longer than in the Wood Sandpiper. The stripe over the eyes is absent in the winter plumage. A rather shy bird. In the Wood Sandpiper the leg colour varies from yellowish-olive, or olive, to brownish-olive with grey joints and the undersides of the wings are pale. The upperparts are dark brown, the feathers having large white spots at the tips in the breeding plumage, but only narrow pale borders in the winter plumage. The breast spotting continues on to the flanks and the tail has 5-8 cross bars. A sociable bird.—**Distribution:** Green Sandpiper has a range in Europe and northern Asia, in marshy country. In Britain it is seen when migrating, mostly in England and Wales. It has nested on 1 or 2 occasions. The Wood Sandpiper has the same breeding range and is also a passage migrant in Britain. It has bred twice in northern Scotland during recent years.—**Diet:** insects, crustaceans, molluscs, small fishes.—**Breeding:** the Green Sandpiper nests in trees, usually in a disused Thrush's nest and the eggs are incubated probably by both sexes. The Wood Sandpiper builds a well-concealed nest on the ground and both sexes take a share in incubating the 4 eggs.

Knot

Calidris canutus

Characteristics: a sociable, compact wader, slightly larger than a Thrush, with a short, straight black bill. In the breeding plumage the upperparts are black with chestnut-brown edges to the feathers, while the head and underparts are chestnut-brown. The legs are dark green. In the winter plumage the upperparts are ash-grey, the underparts white with pale grey stripes on the breast and flanks, and the legs greyish-green. Juveniles are like adults in winter plumage, but with beige flanks and the ash-brown feathers on the upperparts have a dark central streak and black and white edges; the legs are olive-yellow.—**Distribution:** breeds in northern Asia and North America. A winter visitor to Britain, and also seen when on migration.—**Diet:** worms, crustaceans, milluscs and insects. Often seen feeding on the shore in large numbers. On the breeding ground spiders and seeds are also eaten.—**Breeding:** nests on the ground. The 4 eggs are incubated by both sexes for 21-22 days, and the young are fledged at about 18 days.

Purple Sandpiper

Calidris maritima

Characteristics: a compact wader, about the size of a Starling, with a stout, slightly curved orange bill, becoming olive or dark brown towards the tip. The short legs are yellowish-red in the breeding plumage, yellowish to brownish in the winter, and in juveniles. The rump is black with white sides and the tail dark grey. In the winter plumage the upperparts, head, neck and breast are dark grey, the throat and underparts white. In the breeding plumage the grey of the crown, breast and neck is marked with brown spots and streaks. The underside of the wings is white.—**Distribution:** northern Europe, northern Asia, North America, mostly on rocky shores. In Britain seen as a winter visitor and migrant on passage.—**Diet:** molluscs, woodlice, crustaceans, insects, algae.—**Breeding:** the 4 eggs are incubated, mainly by the male, probably for about 21 days.

Little Stint
Calidris minuta
(fig. above)

Temminck's Stint
Calidris temminckii
(fig. below)

The smallest of the waders seen in Europe, only the size of a Robin, and difficult to distinguish. These are sociable birds. —**Characteristics:** legs and bill black in the Little Stint, greenish-olive or olive-brown in Temminck's, but naturally often blackened by mud. The outer tail feathers are grey in the Little Stint, white in Temminck's. The former flies off at a low angle, whereas the latter has a more erratic, twisting flight. Young Temminck's are a uniform brownish-grey above, whereas young Little Stints are blackish-brown. In the breeding plumage Temminck's Stints have two feather types on the upperparts: grey-brown and blackish-brown with ochre-coloured edges. In winter plumage both species are grey above, but Temminck's has a greyish breast, whereas the Little Stint has whitish underparts.—**Distribution:** both species breed in northern Europe and northern Asia, and both occur in Britain as migrants on passage, Temminck's Stint more rarely than the Little Stint.—**Diet:** for both species, insects, worms, small molluscs and crustaceans, seeds.—**Breeding:** both species nest on the ground. In the Little Stint the clutch usually consists of 4 eggs but there is some doubt about whether both parents take part in incubation, and the period is not known. Temminck's Stint also has 4 eggs but here again the details of incubation are not known.

Dunlin

Calidris alpina

(*fig. above* breeding plumage,
below winter plumage)

One of the commonest waders in Europe.—**Characteristics:**
only the Dunlin has a black patch on the lower breast (in the
breeding plumage). The moulting dates vary considerably and
so it is possible to see some birds in winter plumage and some
in breeding plumage at the same time. In the breeding
plumage the back is black with broad, rust-brown edges to the
feathers, the neck and breast are white with brownish-black
streaking. The juvenile plumage is similar to the breeding
plumage, but the ground colour of the neck and upper breast
is grey, and the underparts are white. In the winter plumage
the upperparts are grey-brown, each feather with a thin, dark
central streak, the underparts white, the neck and the sides of
the breast grey with brown streaks. The bill and legs are black,
the bill being longer than the head and slightly decurved.—**Distribution:** Europe, northern Asia, North America,
on moorland and tundra during the breeding season, on
coasts in winter. Dunlins breed in the Orkneys, Shetlands and
Hebrides, and elsewhere in Scotland, and in smaller numbers
in parts of northern England, Norfolk and a few other
localities.—**Diet:** marine worms, molluscs, crustaceans, insects.—**Breeding:** nests on the ground in a well-hidden site,
usually close to water. The clutch of 4 eggs is incubated by
both sexes for 20-22 days, and the young are tended by the
parents for 19-20 days.

Sanderling

(fig. winter plumage)

Calidris alba

Characteristics: a compact wader, about the size of a lark, with black legs and a short, straight bill. In the winter plumage (August to April or May) the upperparts are pale grey, with broad white wing bars and a black rump with white at the sides. In the breeding plumage (April or May to July) the upperparts, breast, neck and head are pale rust-brown and dark brown with white feather tips. The chin, the base of the bill and the underparts are white.—**Distribution:** circumpolar, northern Europe, northern Asia, North America, on tundra during the breeding season, on coasts in winter. A winter visitor and migrant on passage in Britain.—**Diet:** molluscs, marine worms, crustaceans, fish carrion and some plant food.—**Breeding:** nests on the ground. The clutch usually has 4 eggs which are incubated by both sexes for 23-24 days.

Curlew Sandpiper

Calidris ferruginea

Characteristics: a slender, long-legged erect wader, about the size of a Starling, with a white rump and grey-brown tail. The black bill is normally longer and more decurved than that of the Dunlin (p. 104). The legs are black. In the breeding plumage (April-July) the curved bill and the chestnut plumage are unmistakable, the upperparts being blackish-brown with a pattern of rust-brown. In the winter plumage (September-March) the upperparts are grey-brown, the underparts white, the upper breast having beige-grey stripes. During the autumn migration period (July-September) the adults are usually in a transitional plumage.—**Distribution:** breeds in northern Asia. In Britain known as a migrant on passage.—**Diet:** small molluscs and crustaceans, worms, insects, seeds.—**Breeding:** sexually mature in the 2nd year. Nests on the ground, but the exact times of incubation and fledging are not known.

Ruff

Philomachus pugnax

(*fig. above* 2 males in breeding plumage, *below* female)

Characteristics: a silent bird, about the size of a pigeon, the female or Reeve being smaller than the male. Rump white with black in the centre, tail black. Well-known from the characteristic plumage (March or April to June or July) of the males. Their neck ruff and ear tufts are brightly coloured but very variable. The face is naked. In the winter plumage the large blackish-brown feathers have broad whitish to beige edges, the male often having some grey feathers with white edges, and also white on the head or neck. During the same period the chin is white, the breast, and in the male also the flanks, are pale ash-brown, and the underparts are white. Juveniles resemble the female in winter plumage, but the feathers on the upperparts have rusty-yellow edges, those on the underparts ochre-coloured edges. The female in breeding plumage has the upperparts patterned as in the winter plumage, but the neck, breast and flanks are usually pale brown to rust-yellow. The bill and legs vary considerably in coloration.—**Distribution:** Europe and northern Asia. Seen in Britain as a migrant on passage, more frequently in autumn than in spring, mainly in marshes, swamps, wet meadows and sewage farms.—**Diet:** mostly insects, but also worms, molluscs and seeds.—**Breeding:** the males gather in groups of 3-50 on special display grounds which are used year after year. The females come to the grounds, choose a male and mate. Dominant males mate with more females than males lower in the hierarchy or "pecking order". They nest on the ground among grasses and the clutch of 4 eggs is incubated by the female alone. The young are fledged at 25-27 days.

Avocet
Recurvirostra avosetta

Characteristics: a very elegant wader about the size of a crow, with black and white plumage, very long blue-green legs and a long, delicate, upturned bill. The eyes are red or red-brown in the male, brown in the female. In juveniles the black areas are dark brown and the individual feathers have pale rust-brown tips.—**Distribution:** Europe and northern Asia. In Britain the Avocet became extinct as a breeding bird for almost a hundred years. In 1947 several pairs began to nest on Havergate Island in Suffolk where there is now a strong colony. Elsewhere in Britain Avocets occur as rare migrants.—**Diet:** insects, crustaceans, molluscs, worms, small fishes.—**Breeding:** nests on the ground, often in colonies and close to the water. The 4 eggs are incubated by both sexes for 22-24 days and the young are fledged at about 40-42 days.

Black-winged Stilt
Himantopus himantopus

Characteristics: a very slender and graceful wader, about the size of a pigeon, with extremely long pink legs, a black bill and dark red eyes. In the male the back and wings are black, in the female and juveniles in their 1st winter blackish-brown, otherwise the plumage is white. In the male's breeding plumage the back of the head and neck may be black. Juveniles have reddish-grey to dark red legs.—**Distribution:** Europe, Asia, Africa, America, Australasia. Only seen as a rare vagrant in Britain mainly in April-May and September. Two pairs bred in Nottinghamshire in 1945.—**Diet:** insects, crustaceans, molluscs, worms, tadpoles, small fishes and fish spawn.—**Breeding:** nests on the ground in colonies, close to the water. The clutch, usually of 3 or 4 eggs, is incubated by both sexes for about 24-26 days, and the young are fledged at 28-32 days.

Red-necked Phalarope
Phalaropus lobatus
(fig. above)

Grey Phalarope
Phalaropus fulicarius
(fig. below)

The phalaropes are highly specialized waders which swim well and bob up and down on the waves like a cork. The roles are to some extent reversed for the female is more brightly coloured and she courts the male and leaves the care of the brood to him.—**Characteristics:** in the breeding plumage of the Red-necked Phalarope the bill is black, the upperparts and head slate-grey, the underparts and throat white and the sides of the neck orange, the male being much duller than the female, with more buff on the upperparts. In winter plumages the two species are very similar but the Red-necked Phalarope has a longer and more slender bill and a darker grey back with whitish streaks, and it is a slightly smaller bird. The Grey Phalarope in breeding plumage has a yellow bill, whitish sides to the face, a dark brown back streaked with buff, and chest-nut-brown underparts.—**Distribution:** of the Red-necked Phalarope, Europe, northern America, North America in bogs with some open water. In Britain it is seen mainly as a summer visitor, but it breeds in small numbers in the Hebrides, Orkneys and Shetlands. The Grey Phalarope has the same range, but in Britain it does not breed and is recorded mainly as a migrant on passage.—**Diet:** for both species, mainly insects and small crustaceans and molluscs as well as spiders and worms.—**Breeding:** both species nest on the ground, usually in small colonies. In each case the clutch of 4 eggs is incubated by the male alone for about 20 days. The young are tended by the male, although there is evidence that the female may be present. There is only one brood in the year.

Stone Curlew

Burhinus oedicnemus

Characteristics: an extremely shy, secretive yellow-brown bird, about the size of a Wood Pigeon, with large yellow eyes and thick yellow legs. The head is large and broad, the stout bill yellow with a black tip. There are two pale wing bars. The plumage is streaked with dark brown. The bird is mainly active at dusk and during the night.—**Distribution:** Europe, Asia, Africa. Mainly a summer visitor to Britain, but it does breed in several parts of south-eastern England.—**Diet:** chiefly insects, spiders, snails, worms, mice and young birds.—**Breeding:** sexually mature in the 3rd-4th year. Nests on the ground, normally laying 2, occasionally 3 eggs. The clutch is incubated by both sexes for 24-27 days and the young are fledged at about 6 weeks.

Great Skua

Catharacta skua

Characteristics: a large, compact, dark brown predatory gull with short legs and a conspicuous white patch at the base of the main flight feathers (the primaries). The large bill and the legs are black.—**Distribution:** northern Europe, and there are other races in South America and New Zealand. Breeds in fairly large numbers in the Shetland Islands, and less abundantly in the Orkneys, Hebrides and northern Scotland. Elsewhere in Britain it is sometimes seen as a summer visitor.—**Diet:** the eggs and young of sea birds, offal. This and the other skuas obtain part of their food by chasing other sea birds and forcing them to disgorge.—**Breeding:** nests on the ground among grass and usually in colonies. The clutch of 2 eggs is incubated by both parents for 28-30 days. The young are fledged at about 6-7 weeks.

Arctic Skua

Stercorarius parasiticus

Characteristics: about the size of a Black-headed Gull (p. 126), and a brilliant aeronaut. The two central tail feathers are elongated and extend about 10 cm beyond the main part of the tail. The legs are black. There are two colour phases: dark and pale. The dark phase is an almost uniform dark brown, but the pale phase has a whitish neck and underparts and a dark brown crown and upperparts. The flight is not unlike that of a hawk.—**Distribution:** northern Europe, Asia and America. A summer visitor to the coasts of Britain, which breeds in the Shetland Islands, Orkneys, Hebrides and northern Scotland.—**Diet:** birds and their eggs, small mammals, molluscs, insects and carrion. Forces other sea birds to disgorge.—**Breeding:** nests on the ground, usually in colonies. The clutch of 2 eggs is incubated by both sexes for 25-28 days and the young are fledged at 27-33 days.

Long-tailed Skua

Stercorarius longicaudus

Characteristics: the smallest of the skuas with two much elongated central tail feathers, which are up to 25 cm longer than the tail. The crown is black, the upperparts a uniform brown, the underparts white.—**Distribution:** northern Europe, Asia and America. In Britain, recorded as a rather rare migrant, mainly along the coasts of eastern England.—**Diet:** small mammals and birds, carrion and food obtained by forcing gulls and terns to disgorge.—**Breeding:** nests on the ground, the female usually laying 2 eggs. These are incubated by both parents for 23-25 days. The young are fledged at 30-33 days.

Great Black-backed Gull

Larus marinus

Characteristics: a large gull, almost the size of a goose and usually considerably larger than the Lesser Black-backed Gull and Herring Gull. Characterized by the black mantle and flesh-coloured legs. In the winter plumage the head and nape are streaked with brownish-grey. The brown juveniles are difficult to distinguish from those of Herring (p. 120) and Lesser Black-backed Gulls. In the 1st winter plumage the feathers of the upperparts have whitish edges and the tail is white with a broad dark brown terminal bar, the bill is flesh-coloured to black and the underparts pale. In the 2nd winter plumage the head and nape are whitish-grey with a little striping, the tail white with a narrow terminal bar. The 3rd winter plumage is like that of the adult but still with brown streaks on the mantle and an olive-yellow bill with a black tip.—**Distribution:** Europe, North America. Breeds in many coastal areas of England, Scotland and Wales.—**Diet:** fishes, worms, molluscs, young birds and mammals, carrion.—**Breeding:** the 2-3 eggs are incubated by both sexes for 26-28 days and the young are fledged at 7-8 weeks.

Lesser Black-backed Gull

Larus fuscus

Characteristics: smaller than the Great Black-backed and with yellow, instead of flesh-coloured, legs. The mantle is black (Scandinavian races) or dark slate-grey (West Europe). The eyes are pale sulphur-yellow. In the winter plumage the head and back have dark brown streaks.—**Distribution:** northern Europe. Breeds in many areas of Britain. Increasing numbers are wintering near to large cities.—**Diet:** fishes and crustaceans on the shore, and also waste at refuse tips near cities and large towns.—**Breeding:** nests on the ground in colonies. The 2-3 eggs are incubated by both sexes for 26-28 days and the young are fledged at about 5 weeks.

Herring Gull

Larus argentatus

(*fig. above* adult in breeding plumage; *below left* a juvenile, *right* the winter plumage)

Characteristics: the legs are flesh-coloured. The mantle is pale blue-grey, the ends of the wings black with white tips. In the winter plumage the head and neck have grey-brown streaks. The eyes are pale yellow. Juveniles in their 1st plumage cannot be distinguished from those of the Lesser Black-backed (p. 118). The bill is black with a flesh coloured base, the plumage brown with pale brown feather edges on the upperparts, the tail dark brown, the legs dark flesh-coloured and the eyes brown. In the 2nd winter plumage the two species can be distinguished. Herring Gulls have the gull-grey back, the underparts paler, otherwise as in the 1st winter plumage. Eyes pale. Lesser Black-backs, on the other hand, have blackish to brown upperparts, mainly white underparts and head, the breast and flanks with brown flecks; the bill is brownish-yellow with a black tip, the eyes pale brown and the legs are becoming yellow. In the 3rd winter plumage the Herring Gull still has brown streaks on the darker mantle, the tail is white with a dark brown terminal bar (usually broken up), the head and underparts are white with grey markings and the bill is yellowish olive-brown. In the 3rd winter Lesser Black-backed Gulls still have some broken streaks in the dark slate-grey mantle and the end of the tail has brown mottling. In the 4th winter plumage both species may still have some dark mottling on the tail feathers.—**Distribution:** Europe, northern Asia, North America. A common resident breeding bird in Britain.—**Diet:** shore-crabs, cockles and other marine animals. Also carrion and garbage.—**Breeding:** nests on the ground in colonies. The clutch, usually of 3 eggs, is incubated by both parents for 25-27 days. The young are fledged at about 6-7 weeks.

Glaucous Gull

Larus hyperboreus

(*fig. above* breeding plumage; *below* juveniles, 1st winter plumage)

Characteristics: the size of a Great Black-backed Gull (p. 118). The mantle is pale grey and the wing primaries are white, without any black. The legs are flesh-coloured, rarely grey, the eyes pale yellow, the bill as in other gulls (pp. 118-121) yellow with a red spot near the tip of the lower mandible. In the winter plumage the head, neck and breast are striped with grey-brown. Juveniles in the 1st winter plumage are pale brown, including the primaries, with fine streaking. The eyes are dark brown, the legs flesh-coloured and the bill pale flesh-coloured with a black tip. In the 2nd winter plumage the birds appear almost pure white, but with slight brown streaking. The eyes are pale grey-green, yellowish-white or brown. The 3rd winter plumage is like that of the adult but there are still some pale brown stripes on the mantle and a few brown markings on the tail and underparts. The bill is greyish to yellow with a diffuse black area towards the tip. In the 4th winter plumage the bill is like that of the adult and the plumage still has a few brown markings.—**Distribution:** northern Europe and Asia, North America. Occurs in Britain as a winter visitor, usually in smaller numbers between October and April, and more frequently along the east coast than elsewhere.—**Diet:** worms, molluscs, crustaceans, fishes, birds, lemmings, carrion.—**Breeding:** nests in colonies on rocky slopes. The clutch of usually 3 eggs is probably incubated by both parents for 27-28 days. The young are fed by both parents, but the time of fledging is not known.

Common Gull

Larus canus

Characteristics: a medium-sized grey and white gull with greenish-yellow legs and bill (without a red spot). In the winter plumage the crown and nape have grey-brown streaks. The juvenile plumage is dark brown with whitish underparts, and similar to that of young Herring Gulls (p. 120). In general, the Common Gull is rather smaller than the Herring Gull with a more slender bill and the wings project further beyond the end of the tail.—**Distribution:** northern Europe and Asia, north-western North America. In Britain it breeds commonly in most parts of Scotland, but not much in England. The common name is unfortunate as this is not such a common bird as the Herring Gull.—**Diet:** worms, insects, seeds, small birds and mammals, carrion.—**Breeding:** nests on the ground, usually in colonies. The 2-5 eggs are incubated by both sexes for 24-26 days and the young are fledged at about 5 weeks.

Mediterranean Gull

Larus melanocephalus

Characteristics: slightly larger than the Black-headed Gull (p. 126). In the breeding season the head is black and the wings pure white. In the adults the legs are reddish-black, in the 1 year olds reddish-brown and at 2 years pale red or blood-red. The heavily built bill is uniform blood-red or reddish-black in the adults. The winter plumage lacks the black cap, as in the Black-headed Gull. Juveniles have brownish to dark grey upperparts, white underparts, a dark tail bar and a blackish stripe over the eye.—**Distribution:** breeds in south-east Europe and Asia Minor. In Britain this species occurs in small numbers as a regular visitor in spring, autumn and winter, mostly along the south and east coasts.—**Diet:** insects, molluscs, small fishes.—**Breeding:** nests on the ground, in colonies. The 2-3 eggs are incubated for about 23-24 days.

Black-headed Gull

Larus ridibundus

(*fig. above* breeding plumage; *below* winter plumage)

Characteristics: a typical small white gull. The wings have black tips and characteristic broad, white front edges, seen when the bird is in flight. In the breeding plumage the head is chocolate-brown with a narrow white ring round each eye, and the bill and legs are deep red. In the winter plumage the black cap is lost, but there is a dark patch on the ear coverts. In the juveniles the bill is dark and the legs flesh-coloured, the upperparts and wings mottled brown and the head and underparts white.—**Distribution:** northern Europe and Asia. A very common resident breeding bird in Britain, nesting along nearly all the coasts, but also extending far inland (e.g. Derbyshire, Staffordshire).—**Diet:** an opportunist which thrives even in cities where it feeds largely on the waste produced by man. Also feeds on insects, molluscs, small fishes, worms, sometimes mice.—**Breeding:** nests on the ground in colonies. The 2-6 eggs (usually 3) are incubated by both sexes for about 22-27 days and the young are fledged at 5-6 weeks.

Gulls and terns (pp. 114-137)

These form a very uniform group of medium-sized to large birds, whose diet was originally carnivorous and taken from the water, mainly in coastal areas. The feet are webbed. The relationship with the waders is shown in the mottled brown eggs and the mottled downy chicks.

Three families are illustrated in this book: the skuas (Stercorariidae), the gulls (Laridae) and the terns (Sternidae). With their long, pointed wings, slender body and forked tail the terns are born aeronauts which feed primarily by diving down to the sea surface.

Little Gull

Larus minutus

Characteristics: a small gull with dark undersides to the wings, instead of white, and in the adults, no black in the uppersides of the wings. In the breeding plumage (April-August) the head is black and the bill dark red. In the winter plumage the head is white, the crown and ear region dark grey and the bill blackish. Juveniles have white undersides to the wings and a black bar at the end of the tail. In their 1st summer juveniles have some white mixed in the black cap, and still some white under the wings.—**Distribution:** northern Europe and Asia. In Britain quite frequently seen as an autumn and winter visitor, mainly along the east coasts.—**Diet:** worms, crustaceans, molluscs, small fishes and some plant food.—**Breeding:** there are normally 3 eggs which are incubated by both sexes for about 27-28 days. The young are fledged at about 4 weeks.

Kittiwake

Rissa tridactyla

Characteristics: a medium-sized pelagic gull with black legs, a yellow bill and dark brown eyes. The wing tips are black without white markings. In the winter plumage the ear region and the back of the head are grey. Juveniles have grey upperparts with a dark brown band across the wings, a narrow black band at the nape, a dark brown tail bar, white underparts and a black bill. In the 1st summer plumage the bill is greenish-yellow with a brown tip, the legs brown, the tail white and the nape band dark grey.—**Distribution:** northern Europe and Asia, North America. The species breeds at several places on the coasts of Scotland and northern England, and also in the West Country.—**Diet:** worms, crustaceans, molluscs, insects, fishes, grasses, seeds.—**Breeding:** nests on the ledges of sea cliffs, in colonies which may have thousands of pairs. Also known to nest on the ledges of buildings. There are normally 2 eggs which are incubated by both sexes for 26-28 days. The young are fledged at about 5-6 weeks.

Black Tern
Chlidonias nigra

Characteristics: a slender bird, about the size of a Blackbird, with long, pointed wings, a black bill and dark red-brown legs. In the breeding plumage the male has a black head, pale grey underparts and white under the tail. The female is paler, with bluish-grey upperparts and slate-grey underparts. In the winter plumage the forehead and underparts are white, the crown and nape are black and the upperparts grey. Juveniles resemble the adults in winter plumage, but have dark brown upperparts and brownish to flesh-coloured legs.—**Distribution:** northern Europe and Asia, North America, in marshes, rivers and lakes in summer, but seen on coasts in autumn. In Britain seen in spring on migration, mostly in northern England and a few places in Scotland.—**Diet:** water insects, spiders, tadpoles, small fishes and frogs.—**Breeding:** nests are usually a mass of vegetation floating in shallow water. The 3 eggs are incubated mainly by the female for about 17-20 days. The young are fledged at 3-4 weeks.

Caspian Tern
Hydroprogne caspia

Characteristics: a large tern, about the size of a Common Gull (p. 124), with a powerful red bill and black legs. In the winter plumage the crown is white with dark streaks. Juveniles resemble the winter adults, but have spotted brown upperparts, and an orange bill with a brown tip.—**Distribution:** Europe, Asia, America, Africa, Australasia, mainly near the sea, but also on inland waters. Recorded in Britain as a rare vagrant, chiefly in May and June.—**Diet:** mainly fishes up to 10 cm long, also eggs and young birds.—**Breeding:** nests in a slight depression in the sand, often in colonies. The 2-3 eggs are incubated by both sexes for 20-24 days. The young are fledged at 4-5 weeks.

Gull-billed Tern

Gelochelidon nilotica

Characteristics: a stocky tern, about the size of a Black-headed Gull (p. 126) with black bill and legs, and a slightly forked tail. In the winter plumage the upperparts are almost white, the nape with scattered black dots. The ear region is dark grey. Juveniles have pale grey upperparts, the back streaked with brown, the wings and the end of the tail with a few brown markings. The crown and eye region are brownish, the forehead whitish, the bill and legs reddish-brown.—**Distribution:** northern Europe and Asia, Australasia, North and South America. A non-breeding visitor to Britain, mainly in the spring.—**Diet:** crabs, worms, insects, small fishes, tadpoles, frogs, lizards, eggs and small birds, small mammals.—**Breeding:** nests in a hollow in the sand, in colonies. The 3 eggs are incubated by both sexes for 21-23 days. The young are fledged at 35 days.

Sandwich Tern

Thalasseus sandvicensis

Characteristics: a slender tern, about the size of a Black-headed Gull (p. 126) with black legs, a deeply forked tail, a black bill with a yellow tip and a distinctive tuft of black feathers at the back of the head. In the winter plumage the forehead is white, the cap greyish-white and the eye and ear region grey. In juveniles the back, wings and tail are pale grey, densely streaked with dark brown, the cap is black with a yellow-brown border and the bill lacks the yellow tip.—**Distribution:** northern Europe and Asia, North America. Breeds in England in Cumbria, Lancashire, Suffolk, Norfolk, the Farne Islands and a few other places and in several localities in Scotland as far north as the Shetlands.—**Diet:** chiefly fishes, also molluscs and worms.—**Breeding:** nests on the ground in colonies. The 1-2 eggs are incubated by both sexes for 20-24 days and the young are fledged at about 35 days.

Common Tern
Sterna hirundo

Characteristics: easily confused with the Arctic Tern, but the present species has longer red legs and when at rest the tips of the tail feathers do not project beyond the wing tips. In flight the upperparts are greyer and the wings are not translucent. In the breeding plumage the bill is orange-red with a black tip, in winter blackish with a red base. Juveniles resemble the adults in winter, but the back and shoulders are barred with brown.—**Distribution:** Europe, North Africa, Asia, North America. Breeds in several coastal localities in Britain.—**Diet:** worms, crustaceans, molluscs, insects and small fishes.—**Breeding:** nests on sand or shingle, or among short grass, usually in colonies. The clutch of 2-4, but usually 3, eggs is incubated by both sexes for about 21-24 days. The young birds are fledged at about 4 weeks.

Arctic Tern
Sterna paradisaea

Characteristics: this species has shorter red legs than the Common Tern, a blood-red bill usually without a black tip and when standing the tail feathers project beyond the wing tips. In flight the wings appear translucent. In winter the legs and bill are blackish, the forehead white. The juveniles are not distinguishable from those of the Common Tern. —**Distribution:** northern Europe and Asia, North America. Breeds in several coastal areas of England, e.g. the Farne Islands, and also in Scotland where it is more abundant in the north than the Common Tern.—**Diet:** crustaceans, molluscs, insects and small fishes.—**Breeding:** nests on the ground in colonies, near the coast. The clutch of 2-3 eggs is incubated by both parents for 20-22 days and the young are fledged at about 3 weeks.

Little Tern
Sterna albifrons

Characteristics: the smallest of the terns, about the size of a Swift. In breeding plumage this is the only tern with a white forehead. The bill is yellow with a black tip, the legs yellow. In the winter plumage the crown is ash-grey with a brownish tinge, the bill black with some yellow at the base. Juveniles have the forehead sandy-buff, the crown pale brown with blackish streaks, the back and shoulders buff with black feather edges, the tail and rump grey, the legs brownish-yellow and the bill brown with yellow at the base.—**Distribution:** Europe, Asia, N.W. Africa, America, Australasia. Breeds on the coasts in some parts of England, but more abundantly at several localities in Scotland.—**Diet:** crustaceans, molluscs and small fishes.—**Breeding:** nests on sand or shingle, in colonies. The clutch of 2-3 eggs is incubated by both parents for 19-22 days and the young are fledged at 3-4 weeks.

Razorbill
Alca torda

Characteristics: a compact black and white auk, about the size of a duck, with a tall, compressed bill crossed by a central white line. There is also a narrow white line from the upper mandible to the eye. In the winter the chin and cheeks are also white and the white line from the upper mandible is absent. Juveniles have a smaller, pure black bill. The bird moves about erect on land.—**Distribution:** Europe and eastern North America. Breeds on cliffs along most coasts of Britain but not between Yorkshire and the Isle of Wight.—**Diet:** mainly fishes, with some crustaceans, worms and molluscs. —**Breeding:** nests on cliff ledges or among rocks on the shore, usually in small colonies. The clutch consists of a single egg which is incubated by both parents for 33-38 days. When 12-18 days old the young move down to the sea, long before they can fly properly.

Guillemot
Uria aalge

Characteristics: a dark brown and white auk, about the size of a Mallard (p. 50), with a pointed, slender bill without any markings. In the winter plumage the throat and cheeks are white with a black line running from the eye to the ear. The variety known as the Bridled Guillemot has a white ring round the eye and a white line extending back from it. Juveniles resemble adults in winter plumage, with the flanks scarcely striped.—**Distribution:** northern Europe, northern Pacific coasts, eastern North America. Breeds on cliffs round the whole of Scotland and along the coasts of western and southwestern England.—**Diet:** fishes, worms, molluscs and crustaceans.—**Breeding:** nests on the ledges of cliffs, in colonies and often together with Razorbills and Kittiwakes. The single egg, laid on the bare rock, is incubated by both parents for 28-36 days. At 18-25 days the young descend to the sea where the parents tend them.

Auks or Alcae (pp. 136-141)

The auks are relatives of the waders and gulls and they show extreme adaptations to aquatic life. Their uncommonly dense, insulating plumage is well camouflaged when they are in the water, the underparts being white, the upperparts black or dark brown. They swim underwater using the wings, while the webbed feet serve as a rudder. In their adaptations to aquatic life the auks somewhat resemble the penguins, but the two groups are not related. They are underwater hunters which feed almost exclusively on crustaceans and fishes. Apart from the breeding period they spend their time out at sea, usually far from land.

Black Guillemot

Cepphus grylle

Characteristics: in breeding plumage a black auk, about the size of a pigeon, with broad white wing patches and red feet. Flies with rapid, whirring wing beats. In the winter plumage the back has a pattern of white and blackish, the head and underparts are white and the wings are mainly white. Juveniles are similar to adults in winter plumage but are more greyish with some dark grey-brown, the white wing patches are striped with dark brown and the underparts also have dark brown markings.—**Distribution:** northern Europe, Asia and America. In Britain, this species breeds along the western coasts southwards to Cumbria and on the east from the Shetlands to Caithness and Kincardine.—**Diet:** worms, molluscs, crustaceans, small fishes, and seaweed.—**Breeding:** nests in crevices or among boulders, often with a few pairs close together, but not in true colonies. The clutch of 2 eggs is laid on the bare rock and incubated by both sexes for 21-30 days. The young remain in the nest until fully fledged at 36-40 days, and are then quite independent of the parents.

Puffin

Fratercula arctica

Characteristics: an erect-standing, stocky, black and white auk, with orange feet. The unusual bill has areas of grey-blue, yellow and red. The height and number of ridges in the bill is a measure of age. One-year-old birds have one ridge or more. There are two ridges at 2-3 years, three at 4-5 years, and four at 5 years and over. In winter the bright horny parts of the bill are shed and it becomes smaller, narrower and mainly yellow. Juveniles have a dark grey-brown bill, half the height of the adult's.—**Distribution:** northern Europe and eastern North America. Breeds on the coasts of Britain from the Isle of Wight west to Cornwall, and more abundantly along the western coasts to the Shetlands. It also breeds at a few places on the eastern coasts of Scotland, and at Flamborough and the Farne Islands.—**Diet:** mainly small fishes.—**Breeding:** nests in a burrow made by the birds or in one taken over from rabbits or shearwaters. The single egg (rarely 2) is laid about 30 cm from the burrow entrance and is incubated probably by both parents for 40-43 days. The young leave the nest at 40-51 days.

INDEX
English Names

142